Teaching the
Basic Medical Sciences:
Human Biology

Recent Macy Foundation
Publications

Schools of Public Health: Present and Future, edited by John Z. Bowers
and Elizabeth F. Purcell

Medicine and Society in China, edited by John Z. Bowers and Elizabeth
F. Purcell

*Serve the People: Observations on Medicine in the People's Republic
of China,* by Victor W. Sidel and Ruth Sidel

Public Health in the People's Republic of China, edited by Myron E.
Wegman, Tsung-yi Lin, and Elizabeth F. Purcell

Intermediate-Level Health Practitioners, edited by Vernon W. Lippard
and Elizabeth F. Purcell

TEACHING THE
BASIC MEDICAL SCIENCES:
HUMAN BIOLOGY

Report of a Macy Conference

Edited by JOHN Z. BOWERS
and
ELIZABETH F. PURCELL

JOSIAH MACY, JR. FOUNDATION
One Rockefeller Plaza, New York, New York 10020

Contents

Introduction

A strong emphasis on rigorous training in the basic medical sciences has been a keystone of medical education in the United States for the past sixty years. Today, several major forces affecting this traditional pattern of education are creating an acute dilemma for basic medical science departments in their relationship to the educational process that precedes and follows the usual two years of preclinical training.

Biological chemistry and microbiology, for example, have increasingly become the province of general biology. Subject matter that was previously the exclusive responsibility of medical school basic science departments is now presented in courses in "molecular biology" at the college level.

As a result, students are entering medical school with better and more advanced scientific training than in the past, and the degree of sophistication of their premedical education often surpasses the level offered in basic science departments.

A second expanding thrust in the basic medical sciences is coming from the clinical faculty who are incorporating subject matter from the basic sciences in their educational programs. They have the decided advantage of presenting the material with a clinical relevance that is especially appealing to medical students.

A third factor is the sharp reduction in teaching time allotted to basic medical science departments in the new core curricula.

It was the purpose of this conference—held in Williamsburg, Virginia, on November 26 to 29, 1973—to examine trends in education in the biomedical sciences/human biology, within the framework of the university, as they affect premedical students, medical students, and students with other career objectives.

John Z. Bowers, M.D.
President

22 April 1974

Teaching the Basic Medical Sciences:
A Survey of Medical School Curricula

L. THOMPSON BOWLES

Teaching the basic medical sciences is an interesting and complex problem in which all of us share a common interest and involvement. My own perspective is that of a practicing surgeon. I have, however, also had an opportunity to learn the language and some of the techniques of the educator during the past two years while studying comparative medical education in the United States and Canada in behalf of the Association of American Medical Colleges (AAMC). This activity has resulted in two editions of the *AAMC Curriculum Directory.*

The material I shall present was collected during these curricular surveys for the directory. The general purpose was to organize a reliable data base that would facilitate the study and evaluation of medical education. Although the data and their organization still need much refinement, many data-collection pathways and the organizational capability for data management have now been established within the AAMC.

The Basic Sciences in Medicine

Training in the basic sciences has become an increasingly important component of medical education in the United States during the past century. In recent history, Abraham Flexner's Report of 1910 gave emphasis to the necessity for basic sciences in the education of the effective practitioner. The story of medical education in North America since then is well known and I shall not repeat it except to say that much Flexnerian orthodoxy still prevails, and that in most medical schools the sharp division between basic sciences and clinical sciences continues to exist. There is, however, a significant and increasing variety of curricular patterns designed to teach basic sciences to medical students. Within the variability of these educational styles, some strengths and liabilities are identifiable that might well serve as the basis for our discussions.

Many of us have worked with, and accepted referrals from, large numbers of primary care physicians whose principal activity is delivering patient care. While I am continually impressed with the stamina and commitment of most of these professionals, it is apparent that precious little basic science information is utilized in much of their clinical decision making—at a primitive cerebral level some old facts are recalled, much as one summons the proper tense in rapid conversation. Many busy and respected physicians do, however, prescribe penicillin over the telephone for a patient complaining of a sore throat, and in most cases the patient soon recovers, lulling the practitioner into accepting a causal relationship between prescription and patient outcome. While most physicians know that proper medical care calls for at least an examination of the throat and the taking of a throat culture before instituting therapy, many do not follow this more enlightened regimen. Innumerable clinical situations can be cited, not all outside the academic center, to suggest that a great many practicing physicians either forget or lose faith in most of the basic sciences they learned as first- and second-year medical students, and make little effort to renew their knowledge and understanding of basic sciences during their professional careers.

The fact that so many patients get well following visits to their doctors is as much a tribute to the sturdiness of the human biological condition as it is to the wizardry of the healing arts. If this observation is correct, one might suggest that we overemphasize basic sciences in medical education. A case can be made for giving eagle scouts a four-week course in how to use the Merck manual—it is possible that by so doing our primary care system would change very little.

I cannot see how a physician will be able to practice effectively during the coming decades unless he or she develops a continuous learning habit incorporating basic science information and an understanding of concepts that are essential in a doctor's clinical activity. If such an understanding is necessary, how can we organize the curriculum so that medical students will not see their early courses as obstacles to be overcome before they learn to become doctors? How can we demonstrate to them the crucial importance of basic sciences in clinical practice?

In organizing an effective learning program in basic sciences, it must be acknowledged at the outset that the vast majority of entering medical students are not scientists and have few ambitions in that direction, even though large numbers of them have had a considerable amount of biology and chemistry during their baccalaureate years.

Six-Year Programs

Preparation for medical school is a controversial phase of medical education and, while harboring a strong interest in and preference for certain kinds of premedical studies, I shall make only limited reference to that period in this presentation. There are now eleven schools that have grafted the premedical phase onto their medical curriculum.[1] In each case, applications are accepted from high school seniors who begin with the first baccalaureate year and go through a six-year course culminating in the M.D. degree. The six years are organized in a variety of patterns. At the University of Missouri in Kansas City, for example, the entering student switches back and forth from the medical environment to the college campus, spending increasing amounts of time in the hospital as he progresses. Jefferson, Hahnemann, and Albany medical colleges utilize affiliations with Penn State, Wilkes College, and Rensselaer Polytechnic Institute, respectively, to prepare students to enter the basic science part of the medical curriculum. At Michigan, Boston University, Louisiana State in Shreveport, Northwestern, South Carolina, Brown, and Miami, students remain on the university campus for the full six years, the first two consisting of intense premedical science preparation. Almost all of these baccalaureate-medical school articulations are predicated on admitting very bright high school students with strong science capabilities.

The University of Florida in Gainesville shares a program in medical sciences with Florida State University and Florida A&M in Tallahassee.[2] The latter schools select students early in the baccalaureate course and carry them through the first half of the second year of medical school. Of the basic sciences, only systemic pathology is not covered, and the students transfer to Gainesville in the second half of the second year to complete their clinical education.

In two state systems the entering medical student has one year of basic sciences on campus, and then transfers to a clinical campus for the remaining years of his medical education. The University of Illinois has basic science campuses in Champaign-Urbana and Chicago. After a year at one of these, students transfer to a clinical campus in Chicago, Peoria, or Rockford. It is of interest that both basic science schools have very different curricula, as do all three clinical science schools. At the University of Indiana, students pass through a fairly uniform basic science course at one of six branch schools and then transfer to the medical school in Indianapolis for the rest of their education.

A slight modification of the foregoing is utilized by the state systems of Alaska, Montana, and Idaho, none of which have medical schools. This program, which is based at the University of Washington in Seattle, is known as WAMI (the initials of the four states). Students spend part of their first year on their home state campus and then complete their basic science education in Seattle; clinical training may be taken in the home state or in Seattle.

The Three-Year Schools

Nowhere is the time variability for basic sciences more apparent than in the three-year schools. Seventeen medical schools have adopted three years as the length of their regular programs. Basic sciences occupy between forty-four and ninety-five weeks of instruction, and, within that range, hours of instruction vary just as strikingly. Judging from observations in several departments, it is clear that some chairmen still equate learning with the number of hours of faculty exposure. It is equally clear that many chairmen of both basic science and clinical departments gather a measure of security from their students' performance on Part I of the National Board Examinations—as though the scores were predictive of the future physicians' effectiveness.

The Integrated Curriculum

Although an increasing number of schools are using the integrated teaching style for the basic sciences, there is little data to support its superiority over departmental teaching. Twenty-four schools are now using integrated teaching, an increase of three over the 1972–73 academic year.[3] Wayne State has collected a volume of data suggesting that their students have scored higher in National Board Minitests in basic sciences since they have switched to the integrated curriculum. While of great interest, the newness of the experiment makes the data difficult to interpret.

When individual departments organize courses for medical students, there is a tendency to overemphasize the importance of their respective fields. The student, grim and plodding, suffers from factual overkill and forgets what he learns soon after the examination. Some of what he does remember will be quickly invalidated by new studies.

When interdisciplinary committees plan integrated courses, enormous demands are made on the time and energy of faculty members to plan and update course content and to coordinate their lectures. Text material remains largely disciplinary in orientation, so that additional instructional materials must be generated by faculty to supplement the integrated courses.

A final negative note on integration is the interesting fact that a great many basic science faculties are distressed at losing departmental control over both teaching and learning.

There are considerable strengths, however, buried in the tedious and time-consuming integration of material. Most heartening is the fact that students seem to like this system. Departments must communicate with one another, as they have not done in the past. Material is manipulated and learned in the form most resembling clinical decision making. A patient who presents with a urinary tract infection, for example, is an integrated problem, not one concerning just anatomy, or just physiology, or just biochemistry.

We move on from the integrated curriculum to other interesting organizational styles. At Duke an attempt is being made to show the student the entire resources of the school in the first two years: the first year is all basic sciences, and the second all required clerkships. Thereafter the student has one or two years to develop electives in both areas. At Hahnemann the first two years follow a structure similar to that at Duke. Then comes a twenty-four-week segment of systems oriented to basic sciences, followed by a final clinical electives program.

Retention of Knowledge

The academic talents of most entering medical students are such that the format of our teaching can rarely damage them too severely: they usually manage to do what is asked of them. Indeed we might seriously question how important the curriculum really is. Whatever our students learn in the basic sciences, much is quickly forgotten; they subsequently relearn what they need to know in residency training. What they use on a daily basis they come to understand extremely well. If the graduate physician participates in laboratory or clinical investigation, he will be continually reviewing and renewing his knowledge of the basic sciences. Those with the responsibility for teaching medical students and house staff must remain current with advances in the basic sciences relevant to their fields or risk appearing foolish on a regular basis while attempting to respond to the kinds of searching questions only a medical student or intern can think up.

Over 90 per cent of our medical school graduates, however, go on to practice, where they will have neither teaching nor research responsibilities. Within ten years of completion of training most fall into practice patterns that are as much habit as anything else. Unless physicians come to regard related basic sciences as part of their routine reading, it is likely that much of their clinical decision making will reflect mechanical responses to clinical clues rather than rational analyses based on an understanding of human pathophysiology.

If we assume that intelligent people learn what they perceive to be important, and remember information and techniques that they use frequently, the method of teaching medical students may change. Basic scientists have a vocabulary to impart that is essential for an understanding of normal and abnormal human conditions. Moreover, they are especially well qualified to teach the scholarly approach to problems. If all medical students entered their first clerkships possessing a clear vocabulary with which they could describe what they see on the wards, I would be delighted. If they could observe and question the very common failings in clinical decision making they might learn to protect themselves from falling into the clinician's trap. We all know the introductory catch phrases that begin: "Well, in my experience," or "I once had a case." While I hope our students will not respond disrespectfully to us when we fall back on our old clinical supports, I also hope that basic scientists and scientifically competent clinicians will offer exciting alternatives for the analysis of clinical data.

It would not disturb me if one of my students could not recite the anatomic relations around the common duct; if, however, a third-year resident in surgery could not describe that anatomy in detail, I would be worried. How many fourth-year medical students can relate the Henderson-Hesselbach equation to the management of respiratory acidosis? And yet I expect residents in surgery and anesthesia to manage respirators, using blood-gas data that has been rationally analyzed and not twisting dials on the basis of a case they heard about once.

No one medical student will ever know a significant fraction of the basic science information that now exists in anatomy, physiology, biochemistry, microbiology, pharmacology, and pathology. We can only hope that, given time, faculty support, and learning resources, students will learn what they need to know. This process can be facilitated greatly if the faculty has decided what the student should know and has communicated that information to him.

Independent Study Programs

Institutional and course objectives have been developed at Iowa, Ohio State, and Illinois. At the present time, independent study programs are enjoying some popularity at Ohio State, Illinois, and the Medical College of Georgia. These programs are essentially noncurricular: students proceed on their own, in their own time, toward stated objectives, and, utilizing a variety of learning resources, pursue their goals. Data from the latter three schools indicate that students in independent study groups are indistinguishable from other students in their subsequent ward performance and in the test scores on Part I of the National Board Examinations.

In closing, I must admit I do not believe there is much information supporting one style of basic science teaching over another. No method of teaching basic sciences is inexpensive, and I am skeptical of all claims of the economic efficiency of any of the formats I have described. The proximity of basic scientists to clinicians would appear to be educational and healthy for both. I believe that style and format are important only in employing the best resources in any given location. Different resources may well call for different learning programs.

If there are common objectives for learning the basic sciences in medical school, I would cite three as essential:

1. All students should possess a basic science vocabulary sufficient to describe the clinical phenomena observed on the ward or in the clinic.

2. All students should be expert in the scholarly examination and analysis of clinical problems, as measured by the students' ability to solve them.

3. All students should be able to manage their own learning in basic sciences by identifying standard reliable references in the major fields of study.

NOTES

1. L. Thompson Bowles, ed., *1973–1974 AAMC Curriculum Directory* (Washington, D.C.: Association of American Medical Colleges, 1973): p. 280.
2. Ibid., p. 42.
3. Ibid., p. 276.

General Discussion

The high percentage of declared premedical students who take biology courses is impressive. Estimates are: Yale, 20 to 25 per cent; Stanford, 25 to 30 per cent; and Harvard, 25 per cent. Medical schools are attracting very superior students.

The increasing number of early admissions programs in the medical schools has created paranoia among the faculty of college biology departments who fear the premature loss of their students.

When cellular biology, biochemistry, and genetics are offered at the premedical level, they should be presented as pure sciences and not directly oriented toward medicine.

In the premedical, nonclinical setting there are important interactions among the basic sciences; in medical school the chief interactions are among the clinical departments.

The practice of early notification or selection for medical school may have adverse effects on the students; instead of reducing pressure for academic achievement it may only heighten it.

Undergraduate Teaching in Biology:
A General Overview

CYRUS LEVINTHAL

In attempting to describe biology programs in American undergraduate colleges today, it must be recognized that two quite separate developments have taken place in the recent past: 1) during the past ten to fifteen years the content and level of biology courses have changed; and 2) during the past two or three years there has been a change in the way students and their professors and advisors have responded to the large increase in the number of young people who wish to pursue careers in medicine. I will first discuss each of these developments, and then propose some general principles that I believe should be adopted in planning future developments in both premedical and medical education.

Changes Since 1957

In spite of the political activism and turmoil on American campuses, the 1960s was a period of enormous transition in science departments in the arts and sciences units of virtually all universities and colleges in the United States. In large measure the innovations in biology teaching programs reflected developments in the science itself, particularly the increased emphasis on molecular biology.

It has become commonplace to refer to developments in molecular biology over the last twenty-five years as a major scientific revolution. We now know the molecular basis for the storage of genetic information, and we also understand in rather great detail the mechanisms by which this information is transmitted from parent to offspring. Knowledge of the genetic code and the mechanisms of protein synthesis provides a detailed understanding of the consequences of certain genetic modifications that effect alterations in organisms ranging from simple bacterial viruses to man and his complicated diseases.

Many potentially fruitful discoveries have been made so recently that neither their fundamental nor practical impact is yet appreciated. It seems likely, however, that we may soon expect significant advances in our knowledge of the fundamental processes of animal and plant development; the genetic determination of macroscopic structure in higher organisms; and the ways in which cells of these organisms interact with each other to form complex organ systems. Thus it seems likely that many of the problems that have been classified under the general rubric of the structure and function of animals and plants will begin to be understood in terms of their genetic and molecular bases.

Developments in Fundamental Biology

We can trace the genesis of these developments in fundamental biology to experiments that took place in two different settings during the early part of this century. Most of the contributions that led to the advancements in genetics, cytology, and many aspects of cell biology came from academic biology departments, usually separated into zoology and botany. In the area of biochemistry, and much of microbiology, almost all of the contributions were made by faculty who taught in medical schools or who had research appointments in medical research foundations or hospitals.

The development of molecular biology resulted from the union of these two separate traditions: genetics and cytology, from one side, and biochemistry and microbiology, from the other. In this process an essential role was played by those who entered biology from chemistry, physics, and mathematics. The funds invested by the National Institutes of Health in "health-related" research in molecular biology also constituted a significant factor. The application of these recent findings to problems of human health has barely begun, however, and one hopes that such application will be one of the major activities of medical schools during the next several decades. While the actual development of molecular biology in this country and in Europe has taken place in both medical school departments and arts and sciences departments, in every instance advances have been made by people who merged the traditions of both groups.

Changes in the Classroom

As these events were occurring in the research laboratories, the same trends were taking place in the classroom. Genetics has been playing an expanded role in medical school teaching, and biochemistry has increasingly become one of the basic courses in undergraduate college programs. The material previously taught only in graduate school or medical school biochemistry courses is now offered routinely to juniors or sophomores in undergraduate college programs. The trend toward earlier training in biochemistry has been accelerated by the appearance of a small number of widely used textbooks that in a real sense define the nature of the course offerings. This holds true regardless of whether the courses are taken by college juniors or first-year medical students.

I first became involved in teaching freshman biology in response to a student petition at MIT several years ago for a course in comparative anatomy. Although this struck me as an odd request for MIT students to make, we did have the document in which the students petitioned the faculty for such a course. I regarded this action as such a curious phenomenon that I assembled the students who had signed the petition and encouraged them to discuss what they had in mind. It was very clear: they were bored by being taught about RNA and DNA and protein at the *Scientific American* level in every course they were taking. The students did not quite know what else there was, but they had heard about comparative anatomy and figured that this subject covered material far different from that which was being rehashed in their present courses.

The difficulty we all face is that the story we put in cliché forms, such as RNA and DNA and protein, is extremely exciting and we all like to teach it. It is, however, much more fun to teach it to virgins than to old hags who have been through it too many times. I think if one designs a program in which students go through the material once, perhaps a second time at a more detailed level, and even a third time at a still more advanced level—but not the same material at the same level time after time—then it can be accepted with enthusiasm. Thus our reaction was not to introduce a course in comparative anatomy, but to start structuring a core curriculum for the department. This turned out to satisfy the students' objections, even though it was a different response than they had initially suggested.

A Required Common Core

As the science has developed, it has in fact become rather usual for biology departments to offer a coherent sequence of subjects constituting a required core of material on which other courses can be built. A quite common core is one that includes a general introductory course, with a strong but not exclusive emphasis on molecular biology, and courses in genetics, biochemistry, and cell biology. These four courses, in association with one or more laboratory experience, are then regarded as prerequisites for further training in biology, physiology, neurobiology, structure and function of animals, and many other subjects.

The core described here is of course very different from that which was accepted twenty or thirty years ago. At that time there would have been at least one course in botany and one in zoology, as separate subjects, and the entire emphasis would have been much more on descriptive than on analytical biology. In addition, the present biology core requires a number of associated scientific and mathematics courses that normally include at least one year of calculus and one year of physics, as well as organic and physical chemistry.

In many universities the content and level of these core courses introduce into the sophomore and junior college year material previously dealt with in the first year of graduate school or medical school. A partial explanation of these developments is the increase in the number of well-trained molecular biologists on the faculties of undergraduate biology departments, and the changes that have taken place in the nature of biology itself. Similar developments have, however, occurred in many other sciences during the past twenty years. Material that was presented in graduate physics courses in the early 1950s is now taught to undergraduates in their junior and senior years.

Improved High School Training

These changes reflect the improved high school training of students who become involved in science early in their educational experience. Since the first Sputnik flights in 1957 there has been a significant change in the level of the mathematics, physics, and chemistry taught in our high schools. In virtually all college science departments in schools of arts and sciences this trend has led to the earlier introduction of advanced material in undergraduate as well as in graduate courses. Obviously these innovations have made it possible for more advanced graduate courses to keep pace with developments in the sciences themselves; a student receiving a Ph.D. in physics today has in fact studied many of the scientific advances that have taken place over the last twenty years. This had been made possible by the more effective use that has been made of the student's earlier education.

Thus we do not have the option of being as relaxed about the educational process as we were fifteen or twenty years ago. Part of our response to the increased body of knowledge has been to teach courses in a somewhat more intense manner; part of it has been to delete material we no longer think is particularly necessary.

In introductory physics courses we do not spend nearly as much time requiring students to calculate the moment of inertia of doorknobs, not because doorknobs are less common than they were twenty years ago but because we no longer believe that that particular kind of repetitive calculation is necessary for students to become creative physicists. Likewise, in biology programs we no longer require students to remember the details of plant and animal classifications.

A few of these educational innovations turned out to be disastrous, and we are now returning to some of the emphasis of the previous era. The fact remains, however, that there has been an increased emphasis on science and mathematics in the high schools. It is now quite common to find students entering college with advanced placement in calculus—a very rare occurrence in 1957—which leaves them with extra time to proceed to more advanced material in their science courses if they so desire.

Diverse Backgrounds of Graduate Students

All graduate schools of biology have to cope with the problem of dealing with students who enter their programs with very diverse backgrounds. Some students enter with strong training in analytical biology—including the core courses of biochemistry, genetics, and cell biology—and several other subjects, almost invariably including microbial genetics and physiology, and some organ physiology and developmental biology. These students also have had good training in chemistry, physics, and mathematics. Other students come to graduate school strong in descriptive but weak in analytical biology, or with an undergraduate major in one of the physical sciences, but very little biology. There are great differences in the course work these students must take during the first year or two of graduate school. This type of flexibility has not, however, put excessive demands on either the students or the faculty in most departments. Obviously this is one of the reasons why so many of us who teach premedical students in college have been suggesting that greater flexibility in medical school programs is possible, and that it would allow some students to make better use of their college training.

The Sequence in Scientific Education

Starting from high school, the overall sequence in scientific education is for a student to take courses for approximately five or five-and-a-half years and then start laboratory work, generally of the apprenticeship type. At the end of the five-year, post-high school period, most students begin their laboratory experience, working with their professors and other investigators and students in an organized research group. The apprenticeship research period usually proceeds for two or three years, so that the overall time from high school to the granting of the Ph.D. is between seven and nine years.

The dividing point, as far as institutional affiliation is concerned, comes at the end of four years; the dividing point, in terms of what a student really does and how his life functions, comes at the end of five or five-and-a-half years when he has finished most formal course work and starts his laboratory experience. It is obvious that there are many exceptions to this pattern. There are, for example, schools that make a point of trying to get students into a research laboratory when they first start graduate work. These initial laboratory experiences are, however, usually different in kind from the thesis research that starts after a year or so of preliminary work.

The Purpose of Qualifying Examinations

Almost without exception, between the end of the course period, that is, the end of the five or five-and-a-half years, and the beginning of laboratory work, a series of qualifying examinations is held, which in principle determine whether or not the student will continue; in fact, these examinations rarely make such a determination because almost all students pass them—our selection process is good enough so that it is rare that a student has to be dismissed at the point of the qualifying examinations.

On the other hand, the qualifying examinations are in themselves an extremely important part of the educational process, because in large measure they assess what the students study and with how much intensity, as well as what they discuss with each other. Most of us would agree that an important part of a student's education comes not from listening to lectures but from discussions with other students, from problem-solving sessions they hold with their peers—from the whole process of their interaction, which is designed as a major element in their total education. At best the process is one of self-education, not privately, but in groups. I think the primary purpose of the qualifying examinations is to guide the students and help them make effective use of the process of group interaction that takes place during the first year or so of graduate school.

With some exceptions there are few required courses during the first or second year of graduate study. There is, however, a body of information that the students must master, and required sets of problem-solving and laboratory skills that they have to develop. There is also a desired level at which they must be able to take on new problems, and demonstrate, by giving a seminar or writing a paper, that they can in a reasonably short period come up with a concise and clearcut exposition of the kind of material a particular problem demands.

Obviously, I have been describing what I consider to be the ideal graduate program, and not necessarily the program as it applies to all students or all departments. I am not sure that it applies to most students in my own department, although it is certainly the objective toward which we are working. I do not think there is anything unique about biology in this respect: exactly the same pattern applies to chemistry departments, physics departments, and to all other science departments. Furthermore, there is nothing new in this philosophy of graduate education. It is a pattern that graduate schools have been following for at least the last twenty years.

A series of meetings were held over the last two years with representatives from about forty college biology departments to determine what features our departments had in common. It was found that the courses in biochemistry, cell biology, genetics, and microbiology are very similar in these different institutions, due in large measure to the fact that we use a small number of textbooks which tend to define the common courses. In other subjects such as anatomy, physiology, neurobiology, and, to some extent, developmental biology, the differences in course content at these institutions are much greater. There are even greater differences in the material students actually study in college than are evident from the course descriptions. The fact that the courses are available does not necessarily mean that all or even most premedical students take them. In general, the representatives of undergraduate departments feel that students, particularly in the last few years, tend to select a course program on the basis of their estimate of the strategy that will maximize their chances of admission to medical school, rather than on the basis of enjoying a stimulating and useful education.

Premedical Students

The second general change affecting college biology departments has been the increase in the number of premedical students over the past two or three years. The phrase, "the problems of the premeds," produces strong reactions from anyone concerned with undergraduate college education. The stereotype of the typical premed is a bright, hard-working student who is very calculating in his view of the college experience. His strategy in dealing with college is centered around enhancing his chances of being admitted to medical school. As the ratio of applicants to the number of medical school openings increases, we find that many pre-medical students become so concerned with grades that their involvement with substantive material decreases significantly.

Many students have become so competitive that it seriously interferes with their ability to work together with other students. Perhaps even more serious is the fact that many of these very able individuals become extremely conservative in their course selection: they tend to avoid any course in which they risk receiving a grade lower than "A." This means that the students frequently avoid a course for which they are, in fact, well prepared and which might be extremely valuable in their future careers.

The present situation makes the college education experience of those who do go to medical school much less satisfying and less effective than it could be; it is even more serious for those who are not ultimately admitted to medical school. When a college course program is designed primarily to impress medical school admissions committees it is not likely to satisfy either the educational or career objectives of most students.

The nature of the premedical problem is sufficiently well known so that it is not necessary to describe it in great detail. It is, however, useful to consider various possible solutions to the problem, although not, for our purposes, solutions that call for major changes in society. Given the rate at which student objectives seem to be altering in recent years, it is likely that the presently exaggerated premed issue will vanish within a short time. We do, however, have to face the present situation. Even with fewer premedical students, I believe we should still be concerned with improving the overall educational process in the period between high school and the conferring of the medical degree.

The Desirable Objectives

To me the important and desirable objectives are to make this over-all educational sequence more flexible, more intellectually exciting, and more responsive to the societal needs of improved health care delivery, as well as to encourage the application of advances in the basic sciences to important medical and public health problems. Lowering the age at which a student obtains his medical degree does not seem to have a major advantage; whereas reducing the student's competitiveness and increasing his willingness to engage in intellectual exploration do seem to be worth a great deal of effort. So-called "integrated" or "continuous" programs that place the student in an even more rigidly lockstepped schedule appear to be aiming in the wrong direction. Efficiency should not be our major objective; rather we should be aiming at modifications that will give the whole educational process a different flavor. I shall summarize some of the changes I believe could help move us in the desired direction:

1. For medical students, as for graduate students, the division between course work and apprenticeship or clinical clerkship usually takes place about five-and-a-half years after high school graduation, that is, after about one-and-a-half years of medical school. The emphasis should be less on protecting the vested interests of preclinical departments that insist that students take "their" courses, and more on insuring the knowledge, skills, and understanding of students as they start their clinical experience. Perhaps qualifying examinations of the type used in graduate programs would be more useful than many "required" courses in the first year or two of medical school.

2. Undergraduate colleges should be able to design a greater variety of special programs that are interdepartmental in nature, with student evaluation being less dependent on grades in individual courses and more related to the nature of the courses and the student's overall achievements. Some of these changes might cause problems, since not all of the large "required" courses would survive in a system with greater flexibility.

3. Finally, a system of earlier notification by medical school admissions committees would probably improve the educational atmosphere and experience for those students who are accepted, as well as for those who are not.

The point I want to emphasize is that we are dealing with large numbers of students who are bright, who work hard, and whose current educational deficiencies are in large measure due to their response to what they believe is being asked of them. Regardless of the selection procedure, approximately the same individuals will succeed; their educational experience can, however, be modified in a major way by what they perceive as the demands of the selection process.

(The discussion of Dr. Levinthal's paper follows Dr. Manire's presentation on page 24.)

A Basic Science Department in a
Changing Medical Curriculum

G. PHILIP MANIRE

The University of North Carolina School of Medicine was established in Chapel Hill in 1890 as a two-year school. Around the turn of the century the curriculum was expanded, with instruction in the clinical years being given in Raleigh. In about 1910 the university wisely decided that it did not have the resources to build a good four-year medical school, and from then on all of its efforts were devoted to the creation of what they hoped would be a good school of basic sciences in Chapel Hill.

Following World War II, after a great deal of discussion and much controversy concerning its location, the state of North Carolina decided to expand the two-year program to a regular four-year school of medicine on the Chapel Hill campus. A hospital was built, a clinical faculty was recruited, and the first class of forty-eight students was graduated in 1954.

In 1952, while the clinical faculty was being assembled, general agreement was reached that the time was not appropriate to begin major experimentation in medical education. We had a good reputation for the preclinical education of medical students, who were welcomed at major medical schools on the East Coast, and it was deemed desirable to establish clinical departments and laboratories and to expand the basic science departments to serve not only the needs of medical students but a variety of other students of the health professions. The purpose was to change the school from its primary function of a teaching institution to a modern medical center. Consequently, Chapel Hill had a traditional curriculum until the late 1960s.

Beginning in the mid-1960s there was considerable pressure for curricular reform from the student body, as well as from the large influx of faculty members who had come from a variety of schools across the country. Many of the latter had had experience with major curricular revisions in their previous institutions, and they found our curriculum especially subject to modification.

Curriculum Study

With the appointment of a new dean in 1965 who was particularly interested in the process of medical education, an excellent method for studying the curriculum was established, as was a well-staffed Office of Medical Studies. Every member of the faculty interested in curriculum development or revision was recruited to committees that were to carry out a major study of the content and sequence of the course of study. The committees were set up primarily around organ systems, in order to determine priorities and the resources available throughout the medical school in terms of instructional personnel. There was much enthusiasm among the clinical departments, less in the basic science departments. There was, however, no severe reaction toward this operation.

For a period of over two years the studies went on under the direction of an Educational Policy Committee, whose first chairman was a psychiatrist, David Hawkins, a much trusted colleague and a good organizer. The second chairman was a professor of bacteriology and of medicine, a man of educational statesmanship, and, in this case at least, of very considerable patience. We did nothing especially unusual, and our program does not differ in many details from others in the country. One unique aspect, however, is that when the new curriculum was presented to the faculty, there was not a dissenting note against its acceptance.

The present curriculum is based on an interdepartmental, interdisciplinary approach, with a heavy emphasis on organ systems. Thus the first year consists of courses in cell biology, anatomy, microbiology, and general pathology, with considerable emphasis on an introduction to medicine. The second year has twelve courses, primarily on organ systems, and an introduction to medicine and to psychiatry. All courses in the first two years are taught on a truly interdisciplinary basis, and the course directors and faculty are selected by the Educational Policy Committee rather than by departmental chairmen.

The third year is based on five clerkships: medicine, surgery, pediatrics, psychiatry, and gynecology and obstetrics. The fourth year is entirely elective, for each student must, during his medical education, complete the equivalent of seven four-week elective periods, some in other hospitals, some in outlying areas around the state. Well-prepared students can take one or two electives in the basic sciences during the summer prior to their clinical clerkships.

Advantages and Disadvantages

I will now try to enumerate the principal advantages and disadvantages of the curricular revision. The principal advantage is that the process made the faculty think deeply about the curriculum; it provided a good mechanism to carefully examine the curriculum without controversy, to take a close look at requirements, and to establish new priorities. In other words, the process demanded and received the attention of the faculty; it made the faculty study the purpose of examinations and become involved in the system of evaluation; and it provided a means to effect changes on a regular basis.

While there had in the past, in occasional areas, been contact between the clinical and basic science departments and between the various basic science departments, the creation of course committees has given us an excellent opportunity to make appropriate modifications in the curriculum on a continuing basis.

The curriculum provides students with freedom throughout their medical studies, and to some extent recognizes their previous experience. It allows time in the afternoons for remedial or supplementary studies since we have very few classes after 1 o'clock. The elective fourth year has benefited particularly, as students may choose from a wide variety of activities during that period. It has changed examinations from a driving force to a system of year-end evaluations.

There are of course real disadvantages in such a program, particularly for the basic sciences at whose expense most of the changes were made. (I might add that there were very few changes in the third-year clerkships.) The time devoted to contact with basic science faculty has been significantly shortened, and many of the basic scientists see far less of our medical students than they did a few years ago. Even in the fourth year it is a rare student who elects a course in the basic sciences, and those who do almost invariably take pathology. We have shortened the curriculum; fewer hours are spent in classrooms, and, proportionately, the student now spends much more time with the clinical faculty. A few departmental chairmen act as course directors, but there is less identification of the basic sciences. Formerly, every medical student knew the chairman as well as several members of each basic science department. Now very few do.

We also have much difficulty in evaluating students because we disagree on the importance of evaluation.

Interdepartmental courses conducted by the large clinical faculty tend to become too clinically oriented in the first year. I am in the peculiar position of being the course chairman for microbiology as well as chairman of the Department of Bacteriology. Consequently, I monitor very closely the separate courses in cell biology, of which a member of my department is course chairman; immunology, of which another member is chairman; and microbiology, of which I am chairman. Few other chairmen have such an opportunity in their own areas.

The Teaching Faculty

I would now like to give a brief description of our teaching facility, which I cannot say was planned in a logical manner. It so happened that over a period of years, as we developed a new curriculum, concurrently the legislature decided to give us funds for a separate basic science teaching facility composed of lecture halls, conference rooms, and multidisciplinary laboratories to which students are assigned on a permanent basis. The faculty go to this separate building from their offices, research laboratories, or clinics and teach their classes. Each laboratory can accommodate twenty-four students, and in our experience they make remarkably good teaching areas, ideal for impromptu conferences and discussions. In general we try to plan the laboratory work to provoke discussion. We do not teach a large number of techniques, but seek a basis on which one or two faculty members who spend a great deal of time in the laboratories can work closely with the students.

On the negative side, the basic science facility is expensive. If we were given a second chance we would not combine student study carrels and laboratory work space—when two students are in a room they can study; if there are four, they do less studying; if there are more than four, nobody studies. This has certainly proven to be true in our case. Our students attend classes from 8:30 in the morning until 1 o'clock; there are rarely any afternoon classes. Of those four-and-a-half hours, roughly three are spent in conferences or at lectures, which means that the laboratories are used for less than two hours a day. It would have been better to build the laboratories essentially the same way, but provide carrel space for students on a separate floor and use the laboratories in the afternoons for other teaching purposes.

The Microbiology Department

Now a few words about how a department of microbiology looks at the advantages and disadvantages of the curriculum—how we have fared, and have undertaken to provide maximum service to the medical school and to the teaching program. First, we believe that microbiology remains an important element in a physician's education, and that we have the major role in nurturing this concept.

Due to history and accident we are a university department. Our roots are in the arts and sciences; we are a regular part of the graduate school; we cooperate with the allied health sciences; and of course we are the basic science Department of Microbiology for the medical school, where we are located. It is accepted, however, that we also function in a number of other areas. For example, biology has evolved in separate departments of botany, zoology, biochemistry, microbiology. Soon after I came to Chapel Hill, several of us began an effort to create interdepartmental programs for undergraduate and graduate training: outside experts were called upon; studies were made; and reports were submitted. Due to the strong departmental structure and great distrust of the concept, these efforts failed until recently.

I once discussed the relationships of the various biology departments with a distinguished botanist on our faculty. "What you don't understand," he told me, "is that basically botanists are quiet, introverted scientists, whereas zoologists are very aggressive people." I have since recognized the fear expressed in that statement, and I have found at Chapel Hill that botanists fear zoologists; zoologists do not really trust the basic scientists; basic scientists are suspicious of the clinical faculty; and clinicians are often very insecure university professors.

Fortunately, in the early 1960s an interdepartmental curriculum in genetics was initiated, primarily by the basic scientists. Outside funding was obtained, and much assistance was given to the biology departments in their recruitment of faculty and students and in their search for research support. This program helped greatly to dispel the clouds of mistrust. Within a short time it became possible to begin serious discussions on an interdepartmental curriculum in biology, with major participation by zoology, botany, microbiology, and biochemistry. Our department participated in this development, and members of our faculty now teach cell biology courses and give advanced instruction in microbiology for biology majors who wish to specialize in this area.

Ten years ago we began to reorganize our graduate program, with almost complete emphasis on instruction at the research bench level. The most recent American Council of Education survey of graduate departments cited our department as one of eleven on our campus rated as distinguished and strong.

In the medical school itself we have extended our reach in another direction. Two of the first three members of the Division of Infectious Diseases of the Department of Medicine had their basic appointments in bacteriology, and the third has had a joint appointment with our department on a working level for many years. In the mid-1960s our department played a major role in the recruitment of additional faculty for the Division of Infectious Diseases. Each came with a joint appointment, working in our research space, and collaborating closely with other members of the department. A similar relationship is now developing with the new Division of Clinical Immunology.

Thus our interests, and hopefully our influence, now extend to the thriving undergraduate biology curriculum, from which we are drawing an increasing number of our medical students; to the graduate school; to the basic science medical courses in cell biology, immunology, and microbiology; and to the clinical teaching programs in infectious diseases supervised by members of our department.

Accepting these responsibilities—or seizing these opportunities—has allowed us to build solid groups in microbial chemistry, virology, and immunology, with basic scientists and excellent clinicians in each group. The responsibility for cell biology has provided an opportunity for recruitment in areas of biophysics and mammalian cell genetics not usually represented in microbiology. The active participation of clinicians with excellent training in the basic sciences has allowed—and demanded—recruitment in problem areas of molecular biology that we otherwise could not have afforded. I believe it is extremely important to clinical teachers in infectious diseases to live, at least part-time, in such an environment, just as one of the primary functions of basic science departments is the continuing education of the clinical faculty.

We are developing a model system for the growth and survival of basic science departments in medical schools. I believe an education in the basic sciences is more essential for the young physician than it has ever been. It is very difficult for a medical school to achieve excellence without a close relationship with an equally good university, but, in the long run, no medical school is likely to be better than the university with which it is affiliated. It is equally true that no medical school can have a strong clinical faculty if it does not have a strong basic science group. I have observed that in well-respected schools with weak basic science departments, the clinical departments usually have large numbers of non-clinical basic scientists on their faculties. I would say that even though the basic scientists have no contact with medical students—a situation I hope never to see—a good medical school should of necessity have strong basic science groups.

It is a fact, however, as I mentioned earlier, that curricular revisions in medical schools have been made primarily at the expense of the basic sciences, or at least of basic science departments. As times get harder and funding more difficult there is going to be an increasing tendency on the part of deans and vice presidents to recognize teaching loads as a prime consideration in the allocation of resources. If this should happen, and if basic scientists participate less and less in medical education, there will be increasing pressures to remove their departments from the medical schools. I suspect this trend will grow on a nationwide basis.

I think we have countered this move at Chapel Hill, at least in our department. Our type of organization safeguards the field of microbiology; provides the interface for scholars with diverse interests to work and learn together and from each other; makes possible many contacts between our faculty and students throughout their medical education and beyond; serves as a good quality control for appointments in clinical areas, since we participate actively in their recruitment; and offers opportunities to relate clinical problems to the basic sciences. This concept allows our department to carry out research on important problems; provide maximum service to the university community; and monitor on a continuing basis the teaching of microbiology to undergraduates, medical students, and postgraduate physicians.

General Discussion

Two provocative challenges were enunciated: (1) transfer large numbers of faculty and students from the medical schools to university undergraduate departments for education in the basic medical sciences; and (2) merge the premedical and the basic sciences so that the total span may be reduced from eight to seven years.

If the basic sciences are moved out of the medical schools, there is a danger that the schools will become mere trade institutes. Moreover, at the baccalaureate level there is the hazard that biology departments may be reoriented and become sharply focused premedical units, thus jeopardizing their broader scholarship function.

Consideration might be given to creating a division of biomedical sciences that would, under joint jurisdiction, embrace both undergraduate biology and medical school basic science departments.

The clinical departments in the medical schools need to associate with the basic science departments more than the latter need the clinicians.

Basic science departments make major contributions in evaluating faculty appointments and promotions; curriculum development and evaluation; collaborative research; and the development of multidisciplinary centers such as those for research in cancer, cardiovascular disease, and genetics.

The approach of the medical school faculty to teaching courses in, for example, microbiology, may be very different from that of the undergraduate faculty; the medical school approach should be available to undergraduates who do not plan to enter medical school.

In establishing integrated efforts between basic science departments in the medical schools and undergraduate biology departments, structural and jurisdictional questions should not be paramount.

The primary concern of medical students in choosing electives is their effectiveness in preparing students to practice medicine.

Human Biology: An Integrating Component in Undergraduate and Professional Education

THOMAS B. ROOS

A Conceptual Basis for a Program in Human Biology

Undergraduate education in the United States, by tradition, rests on a series of assumptions, chief among them being the value of preprofessional training, or preparation for specific careers. This underlies the concept of the multiversity, the strength of academic departmental majors and departments, and the defensive antiscientific attitude of self-styled humanists. Contemporary participants in education seem unable to recognize the simultaneous need for vision as well as competence in scholarship. While factual mastery alone may masquerade as wisdom, it cannot forever disguise a lack of ideas; conversely, a fertile imagination cannot satisfy its promise without a substrate of reliable information on which to act. It therefore seems appropriate to reopen the issue of programmatic instruction, not as an alternative to disciplined study, but as a supplement to it in the interests of liberal education.

Human biology is a particularly apt subject to explore in seeking to find a way to return to liberal education. Indeed what better object exists for a humanistic enterprise than mankind? But, as Alexander Pope knew, the way to understand man is to study him directly, not in the abstract. As man is a biological creature, human study, no less than all other aspects of biology, requires an understanding of the relationships between biological systems at levels of organization from molecules to societies: alone, interacting, and changing. If man cannot understand himself by scanning God, neither can he be understood while ignoring his own nature.

A program in human biology would provide a chance for students to apply their special disciplinary interests and talents in their search to understand themselves, as well as mankind. It would allow an illumination of humanistic speculation by the often harsh light of testable observation. Given the importance of language, literature, ideas, and culture

26

to human activity, the biology of that most human attribute, the mind, demands more than classical zoology, psychology, or anthropology can contribute. Rejecting with Lucretius the Cartesian demon of duality, human actions can be examined safely only from a sure awareness of biological capacities and limits.

A program in human biology, undertaken as a supplement to a major, could thus provide a framework within which students might direct their activities from several technical specialities toward a common goal of understanding. It is therefore an example of a coherent program achieved through diversity, rather than a uniform fragmentation derived from professional hyperspecialization. It demands disciplinary excellence, but directs that competence outward as well as inward.

Finally, and politically, a program in human biology that stresses specific disciplined knowledge would not supplant legitimate academic specialities. Its only intended victims are the shibboleths of formless "distributional" requirements and the dishonest practice of asking students to integrate their thoughts and knowledge in ways their teachers avoid. It would facilitate the exchange of factual knowledge that has so often enlivened old disciplines and started new ones.

Possibilities for Students

Human biology might be presented in either of two ways: as a brief introduction to human physiology, or as a serious attempt to integrate contemporary knowledge. The first is easily achievable and obvious in its operation; I shall not discuss it further. The second is more difficult to present and to implement. It should not be taken as a mere euphemism for premedical training: indeed a large part of its clientele should be persons who do not seek careers in any of the health sciences, including medicine. The program must allow each of its participants, whether student or faculty, full freedom to develop and practice his or her individual discipline according to the dictates of personal interest. For students contemplating careers in urban or rural planning, public health, environmental or health law, consumer advocacy, or the traditional social and behavioral sciences, as well as for those intending to practice medicine, prebaccalaureate study in human biology can give practical knowledge of direct future value.

All three major levels of biological organization must be included in the program: infraorganismal (molecular, cellular, tissue, and organ), organismal (individuals as organisms), and supraorganismal levels are common parts of preparatory studies for medical school. The way in which organismal and supraorganismal levels are treated, however, will distinguish a true human biology program from a premedical curriculum. Studies at the individual level must include man's thought and art, as well as his anatomy and physiology. Attention above the level of the individual must be directed both toward his genetic relationships as a member of a breeding population, and his social relationships as a member of political and ecological communities. At all levels the flow of both energy and information must be followed as they enter the system, are used by the system, and ultimately emerge as waste or work from the system. In all cases the growth and development of individual systems must be considered along with, and in contrast to, the evolution of classes of systems.

The program could offer a special opportunity for highly competent students to escape the artificial atomization of social contact by sharing the language and hypotheses of several disciplines. Even were there no practical gain for a lawyer in understanding biological interactions of drugs with organs, or of pollutants with ecosystems, there would still be a reward in the potential for greater intellectual fellowship. The benefit to society of having nonprofessionals conversant with, and able to write about, technological applications of science cannot be overstated. Whether through widening the range of metaphor available for poetry, or reducing the gullibility of journalists towards "scientific" fads or hyperbole, wider dissemination of biological understanding would help overcome the obscenity of the "two cultures."

A Practical Proposal

Two constraints affect any proposal for a serious program in human biology: it must have sufficient content to give its students real knowledge rather than a mere appreciation of human biology, and it must be sufficiently nonrestrictive as to allow participation by students who will neither major in the sciences nor pursue careers in the health sciences. Applying these constraints precludes the renaming of a premedical program as human biology; it demands a new look at courses now available to undergraduates.

The program must include a group of courses planned with regard for each other's content, yet not so homogeneous as to lack cross-disciplinary value. They must be independent enough to permit students to take some without taking all, yet sufficiently related to avoid giving only a glossy finish to shallow comprehension. The program should include both introductory and intermediate courses with minimal prerequisites. All courses should be open to any student, whether or not he or she is a formal participant in the program. Advanced courses should be minimized, since their specialization places them more effectively in the curricula of the various professional schools to which graduates of the proposed liberal arts program may go.

Four areas should be included among the introductory courses: mathematics and logic, the behavioral sciences, literature, and the biological sciences. The mathematics course—one or two terms—should include an introduction to probability and statistics and the logic of linear algebras and computers. Ideally, while the course would permit access to intermediate courses in logic, statistics, and computer science, it should be self-sufficient. (It should not include calculus.) The two-term course in the behavioral sciences should attempt a synthesis of anthropology, psychology, sociology, and social history, but it could probably begin as a pair of introductory courses in any two of these subject areas. In literature almost any course emphasizing the human contemplation of human nature might suffice, whether it takes the form of philosophy, poetry, essay, or fiction. The course in the biological sciences should take one or two terms and emphasize genetics, regulatory networks, and biological time relationships in evolution and development. It must explore the interactions between structure and function and between energy and information in the adapting, developing, and growing individual. Faculty members responsible for the several courses should try to cross-reference each other's material to avoid unnecessary redundancy—in discussing evolution, learning, or philosophy, for example.

The intermediate course should be in "systemic biology"—emphasizing human examples but not restricted to them—and include both laboratory work and seminars as well as lectures (See Appendix A). The term systemic biology is meant to imply an integration of the structure, function, and development of systems and subsystems in vertebrates—especially the human. The course should be planned to follow and make use of all the introductory studies in the program. Thus it would be appropriate when discussing the nervous system to relate its anatomy and physiology to its psychological correlates, its relation to other information-processing systems, and its use in language.

Beyond systemic biology, in schools where conditions permit, an advanced course in human biology should be available as an elective supplement to the program. The advanced course lies outside the bounds of the human biology program, although it might be part of a larger human biology major. This apparent anomaly emphasizes the role of the program as a *preparation* for studying human biology. This preparatory and open-ended nature of the curriculum is important for the liberation of students from the blinders of early professionalism and the fictions of oversimplification.

The advanced course preferably should be taught collaboratively by some faculty members who are engaged in practical applications of human biology, and by others who are pursuing it as a scholarly discipline. Collaboration of this kind would probably require that advanced human biology be taught in part, at least, by the faculties of professional schools of the health sciences; it would thus be limited to colleges affiliated with or located near such schools. Systemic biology and the other components of the program would be available in many colleges independent of professional school associations.

Figure 1 illustrates the relationships among the courses in the program in human biology, per se, and to those courses traditionally required for admission to professional schools of the health sciences. It further suggests where an advanced course might be placed in a more extensive program. A student could gain access to most courses in the human biology program with only one prerequisite. The apparent entrance requirements for the courses in systemic biology merely free it from the constraints of elementary exposition.

As a final caveat, perhaps too obvious to require explicit statement, the success of the program would depend on the strength of the disciplines that comprise it: no program in human biology can survive without a strong and independent base in the humanities, social sciences, physical sciences, and biological sciences.

Further Advantages of the Program

Despite its being distinctly different from a premedical curriculum, a program in human biology would mesh with premedical and medical education in a fruitful and economical way. It would obviously supplement students' preparation for careers in the health sciences. Moreover, with reasonable care in planning and execution, a two-year sequence including systemic biology and advanced human biology could stand in place of such medical school courses as anatomy, physiology, and histology. If supplemented by another two-year course in cell biology, including microbiology, biochemistry, and cytology, and based on preparation in chemistry, physics, molecular biology, and calculus, it could provide a practical alternative to students for their first year of medical school. The same two "macrocourses" would also give the solid background in biology needed for such careers as health care management and medical

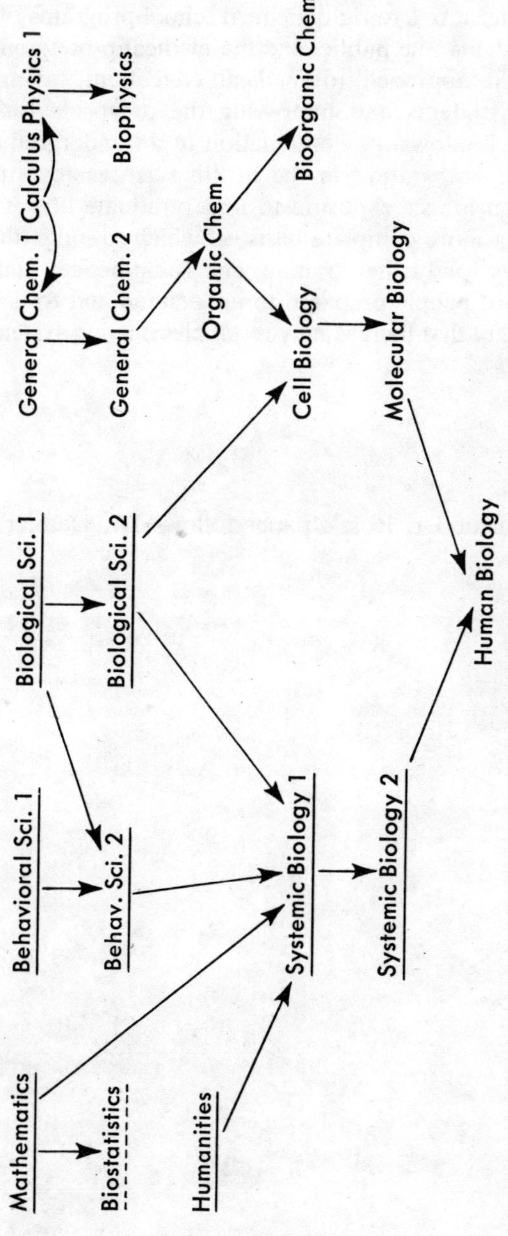

Figure 1. Schematic representation of a program in human biology and some related courses*

* Courses required for the program are underlined

law, without demanding enrollment of nonmedical students in a medical school.

The possible alternative to rigid medical school programs offers several advantages to students, the public, and the medical profession. It permits a more diversified approach to medical education, saving time and tuition for some students and improving the prospects for a medical career for others; it allows the combination of an undergraduate experience with specific preparation in the health sciences for students who would benefit from longer exposure to undergraduate life; it offers professional schools a more complete basis on which to judge the qualifications of applicants for further training and for practice; and, finally, it expands the pool of people prepared to understand and to help maintain an integrated society that increasingly resembles a living system.

(The discussion of Dr. Ross's paper follows Dr. Dethier's presentation on page 42.)

A Proposed Course in "Systemic Biology"

THOMAS B. ROOS

A. GENERAL ORGANIZATION
1. Lectures: About 100 hours
   a) Frequency: Three a week
2. Laboratories: Two hours per hour of lectures
   a) Frequency: One or two a week, depending on the illustrative requirements of the lectures and the availability of material for discussion
   b) Tightly integrated to lecture topics, preferably combining anatomical and physiological observations
3. Discussions: One hour per four hours of laboratory
   a) Frequency: No more than once a week in place of laboratory
   b) Held only when sufficient laboratory experience is available for discussion of personally obtained information

B. COURSE OUTLINE
1. Body form and function (6 per cent)
   a) Adaptive variation
   b) Polymorphism
      i. Sexual
      ii. Maturational
   c) Individual variability of phenotype
2. Integrative systems
   a) Nervous systems (16 per cent)
      i. Microstructure and function: cell types, synapses, and transmission
      ii. Spinal cord
      iii. Brainstem
      iv. Hypothalamus
      v. Thalamus and epithalamus
      vi. Metencephalon
      vii. Telencephalon

       viii. Afferent nerves
        ix. Efferent nerves: voluntary
         x. Autonomic nerves
   b) Chemoregulatory systems (10 per cent)
         i. Neurosecretory and neuroendocrine
       ii. Trophic: general growth and target-organ growth
      iii. Reproductive
      iv. Metabolic
       v. Homeostatic (adaptive)
   c) Sensory systems: afferents (5 per cent)
        i. Electrochemical
      ii. Acoustico-mechanical
     iii. Chemical
   d) Communication systems: efferents (6 per cent)
        i. Ectocrine glands and pheromones
      ii. Light and electric organs
     iii. Epidermal display
     iv. Vocal organs
3. Vegetative Systems
   a) Transport systems (6 per cent)
        i. Vascular: heart and blood vessels
      ii. Lymphatics and interstitium
   b) Respiratory system (3 per cent)
        i. Gaseous exchange
      ii. Gas transport
   c) Blood system (7 per cent)
        i. Plasma
      ii. Formed elements
     iii. Immune system
   d) Digestive system (8 per cent)
        i. Gut tube and small glands
      ii. Enzyme-secreting large glands
     iii. Liver
   e) Purification systems (6 per cent)
        i. Kidney
      ii. Sweat glands
     iii. Reticulo-endothelial system
   f) Storage systems (3 per cent)
        i. Adipose tissues
      ii. Glycogen stores

iii. Protein stores: gluconeogenesis
iv. Mineral storage: bone

4. Reproductive systems (10 per cent)
   a) Gonads and gametogenesis
   b) Gonoducts
   c) Placentation
   d) Gestation and postnatal care

5. Skeletal and locomotory systems
   a) Integument (4 per cent)
      i. Epidermis
      ii. Dermis
      iii. Dermal bone and cranium
   b) Axial systems (3 per cent)
      i. Notochord
      ii. Vertebrae and head articulation
      iii. Axial musculature
   c) Appendicular systems (9 per cent)
      i. Girdles and axial attachments
      ii. Limbs
      iii. Hands and feet

Education in Physiology and Anatomy: Structure and Function of Organ Systems

VINCENT G. DETHIER

Given the nature of the overall topic—teaching the basic medical sciences—it follows that interest in the teaching of physiology and anatomy in undergraduate colleges is stimulated here by the relevance of these subjects to medical education. On the other hand, since the teaching of the structure and function of organ systems is not directed solely to the needs of premedical students, it can be discussed meaningfully only in the broader context of education in the life sciences. I therefore propose, first, to scrutinize the philosophies and methods involved in the presentation of physiology and anatomy in this broad context, and then to address myself to two more sharply delineated issues: the need and desirability of specific courses in physiology and anatomy for premedical students; and the place of human physiology and anatomy in the undergraduate curriculum.

It has been stated that American medical students come from 750 colleges in the United States, and that 30 per cent come from twenty-five of these schools. Obviously, while not all schools offer the same type and quality of training, the majority of premedical students have advanced through some kind of a biological science department. It is therefore reasonable to inquire into the place of anatomy and physiology in the organization and philosophy of these departments.

It has been estimated that approximately 2.5 million high school students enroll in formal biology courses each year. Approximately 200,000 study some form of life science in college, and probably 25,000 major in one of them. Of students who do not major in science, more elect biology than any other field. Biology departments therefore provide many collegians with the only contact with science that they will ever experience. These departments also train the 80,000 professional biologists who will become teachers and research workers, as well as premedical, predental, preveterinary, and preprofessional students destined for careers in health-related sciences.

It is against this background of multiple goals that the curricula of departments of life sciences must function productively. Their approaches to different subjects vary from one institution to another according to the particular view of its mission that each department entertains; the skills and interests of the faculty; local tradition; existing physical facilities; and financial status.

Whereas the meteoric ascendency of molecular biology and the cohesiveness of the biochemical approach have engendered in American education a rather general agreement in approach and an academic universality, no such uniformity can be discerned in the teaching of physiology and anatomy. As one views the academic scene in the United States panoramically, one cannot help but be struck by its tremendous heterogeneity. Historically the subject matter of biology was subdivided into botany, zoology, histology, cytology, embryology, comparative anatomy, genetics, vertebrate physiology, and a variety of phylogenetically oriented courses such as entomology, ornithology, and mammalogy.

In the era between the two world wars the differences in the curricula of the small liberal arts colleges, the larger private universities, and the land-grant institutions were those of quantity and quality rather than of factual content: offerings were in general very similar. With time, some disciplines acquired the status of separate departments. While there was a tendency for botany and zoology to merge into a single department, if they had not already done so, biology tended to fragment into many units. This is hardly surprising considering the almost infinitely broad purview of phenomena related to life. The trend was nowhere more fully realized than in the large land-grant institutions. Physiology and, to a lesser extent, anatomy were two other subjects that achieved departmental status in many institutions.

As biology evolved, as it began to achieve greater unity, room had to be made for the new concepts that were developing, and this began to occur at the expense of traditional courses. It was probably easier to incorporate new knowledge—without sacrificing old—into the existing organization of the large land-grant institutions where the life sciences were already fragmented into many departments: bacteriology, plant physiology, entomology, genetics, anatomy, etc. Other institutions were more restricted by limitations on the size of their faculties, by lack of facilities, and by financial restraints. A college with a total biological science faculty of twenty could hardly hope to offer the same smörgåsbord of courses as an institution that had five or more discrete departments of ten members each.

Then, too, there was a limitation on the students' intellectual absorptive capacities. As the years passed, biology students learned less phylogeny in order to learn more molecular biology. Those in institutions with multiple departments specialized in separate disciplines early in their undergraduate careers. Moreover, as the quality of instruction in biology improved in the secondary schools, colleges began to take for granted the acquisition at that level of the more traditional branches of knowledge. Ironically, the secondary schools were just as anxious as the colleges to teach "modern" biology, and, as a result, subjects such as anatomy and physiology received an ever-decreasing share of attention.

New Courses for Old

As the frontiers of biology advanced, new courses were substituted for the traditional ones. Disappearing from the catalogs were taxonomically oriented courses, comparative anatomy, histology, and traditional vertebrate physiology. In their places appeared molecular biology, neurophysiology, animal behavior, population biology, and others. Remaining from the traditional curricula, although frequently masquerading under new guises, were such fields as genetics, embryology, and the new cytology. In biology departments the character of the change was dictated by the interests of the faculty in residence; their concept of what was important in biology; and their view of the mission of the department.

Although a very large proportion of the students majoring in biology in some institutions were premedical candidates, their special needs seldom if ever dictated the course of development of the departments. A perceived obligation to provide courses that were required for admission to medical school was often met grudgingly. "Service" courses were retained in the catalogs only as long as professional schools demanded that their entrants take them. Thus, for example, comparative vertebrate anatomy continued to be offered in many institutions only as long as it was a premedical requirement; when it was no longer necessary, and as the supply of faculty uniquely qualified to teach anatomy dwindled, the course was dropped with unseemly haste to make room for newly evolved subjects.

Most departments of biology were prepared to drop comparative anatomy *as such* long before the medical schools removed it as a requirement, which indicates that many biology departments perceived their role in education from a different perspective than the medical school in terms of how it envisioned biological training for its prospective students.

The Aims of Biology and Medicine

The aims of biology and of medicine are different, but not by any means mutually exclusive. Medicine is but part of biology—it is even just one part of human biology. Medical schools are interested in one animal, man; whatever their philosophies may be, their energies are directed primarily toward the sick person even though their ultimate goal is preservation of health. The physician is interested in the human species to the same degree, although not with the same ulterior motives as the entomologist whose concern is the insect. Man and insect are only special cases of life—two variations on a single theme. It is possible to acquire a comprehensive and insightful view of life without knowing very much about insects or men; it is not possible to have a profound knowledge of insects or men without a comprehensive knowledge of life.

Insofar as man is an animal, the basic problems of life are common to him: the department of biology is interested in life. This is the interface where medical schools and biology departments meet. The latter deal with generalities for their own sake, generalities that are also basic to the concerns of medical schools for parochial reasons. If the medical school is interested in human physiology and human anatomy, it is not *necessarily* to the advantage of the biology department to offer special courses in these subjects. A sharp distinction must be made between the acquisition of knowledge of physiology and of anatomy and the presentation of such courses. As the Committee on Research in the Life Sciences of the National Academy of Sciences states, there is now general agreement that

> . . . there exists, in the intellectual content of biology, a common core of material that should form the basis of an undergraduate major, appropriate regardless of subsequent fields of specialization. Thus, the same set of courses can serve for the premedical student and for the student who intends a research career.[1]

The Princeton Model

Many departments do not consider that *courses* in vertebrate physiology and anatomy constitute part of the core. At several institutions courses are designed around levels of organization—molecular, cellular, organismic, and population—and numerous variants of this design. At

Princeton, for example, the levels are molecular, developmental, neuro-biological-behavioral, and ecological. In few if any of the curricula designed around levels of organization are anatomy or vertebrate physiology taught as such.

There are of course ways to provide premedical students with the kind of physiological and morphological knowledge that medical schools want them to acquire without the formality of specific courses in these subjects, and without placing a department in the position of having to offer service courses it feels are not consonant with its general philosophy.

Only about one-third of the registered premedical students at Princeton are enrolled in the biology department. Of the majors in that department slightly less than 100 per cent are premedical students. Of all of the undergraduates who apply to medical schools 80 per cent are accepted. While none of these students have had a formal course in anatomy or vertebrate physiology, they have acquired knowledge within a different framework. It thus appears possible to educate students in the realms of physiology and anatomy without offering formally designed courses in these subjects.

In order to illustrate this point let us consider physiology. There are innumerable approaches to the study of physiology: it can be taught by employing one organism, man, for example, as a unifying concept for all physiological phenomena; it can be taught from a comparative viewpoint, system by system; or in a way that emphasizes all-pervasive phenomena as, for example, homeostasis, coordination, or behavior.

The latter approach is the one followed at Princeton. Within the framework of population biology, developmental biology, neurobiology-behavior, and cell biology, students acquire much of the knowledge traditionally presented in a classic physiology course. Neurobiology-behavior, for example, covers such phenomena as neuroanatomy, neurophysiology, neuromuscular physiology, endocrine control, and homeostatic mechanisms; renal, respiratory, and circulatory physiology and the relevant anatomy are presented in the introductory course; cellular physiology is treated in cell biology; events at the molecular level in the biochemistry department; and adaptive and evolutionary aspects of physiology are treated in behavior and ecology. There are no courses in human physiology or anatomy, as such.

It would appear from the results of medical school applications that this approach to education in the structure and function of organisms satisfies—or at least does not frustrate—the needs of premedical students, while harmonizing with the presentation of biology for those who will become professional biologists. At the same time it fills the function of providing a liberal education. The system is, however, generally less appealing to medically oriented students and to some professionals who think pragmatically in terms of specific and immediate goals and who are acutely conscious of a need to utilize time with maximum efficiency.

"Human" vs. "Humanistic" Biology

In many quarters a drive is underway to institute courses in human biology, but the various advocates describe the subject in different terms. Furthermore, the motives range from a desire to accelerate medical education to a recognition that there exists in this generation a fresh and compelling awareness of man's integral place in nature. It might help to clarify the issues by differentiating between "human" biology and "humanistic" biology.

Pedagogically, the desire to offer students instruction in human biology can be fulfilled by such courses as human physiology, human anatomy, human parasitology, human genetics, immunology, and pharmacology. While these courses might satisfy the particular needs of premedical students, they less satisfactorily assuage the appetites of other students of biology. For one thing, they particularize; for another, man is only occasionally a good model for things biological. Where he is the most suitable model, man should by all means be used, and, indeed, in thoughtful courses in development, cell biology, neurobiology, and population biology, he receives his just due. To design special courses around man, however, especially in relation to structure and function, is to place undue emphasis on technical information—a luxury most departments of biology can neither afford nor deem desirable.

Humanistic biology offers a less restricted horizon. The distinction is not merely one of semantics; it is based upon viewpoint and goal. Humanistic biology can take many forms: it can be a curriculum that collectively scrutinizes man as a unique biological, psychological, and social phenomenon, without concentrating individually on the human animal; it can be a single course that covers the species in the context of the total biological and physical environment; or it can introduce man into courses at those precise points where he provides the best example of, or conversely, a unique exception to, the phenomenon being studied.

It is in this last context that structure and function can be taught successfully in such a way as to *provide* premedical students with knowledge, insights, and comprehension that will establish a foundation for their clinical years, while simultaneously constituting one element of a liberal education for them, as well as for biology students. Our parochial experience has been that in trying to teach the best *biology* we provide first-rate premedical training, and that in presenting physiology and anatomy as integral components of other conceptual approaches to the life sciences we satisfy divers needs. At the same time, with reciprocal neighborliness we are able to cross departmental fences with impunity and enrich our program with contributions from engineering and psychology.

There is no best way to provide education in physiology and anatomy —each institution must approach this goal in the manner most suited to its own structure and function. However this is achieved it is wise to ponder the words of Bentley Glass before deciding that specific courses in vertebrate or human physiology or anatomy in the traditional mold— albeit modernized—are either necessary or desirable elements of premedical education in the colleges. Glass's thoughts may be paraphrased as follows:

> Science is not *just* bodies of knowledge—facts and concepts. The increase of power, once haphazard, empirical, and painfully slow, is now frighteningly swift through the growth of the natural sciences and the application of scientific knowledge in engineering, medicine, and agriculture. The essence of the idea of a liberal, as opposed to a technical, education is freedom—freedom of the mind that grows out of an indomitable quest for truth. Enlargement of freedom by *empirical* science is by means of the extension of choice and opportunity. Freedom of mind comes from pure science.[2]

In electing how it shall present physiology and anatomy the average biology department is conscious of these broad responsibilities.

NOTES

1. Committee on Research in the Life Sciences, *The Life Sciences* (Washington, D.C.: National Academy of Sciences, 1970).

2. Bentley Glass, *Science and Liberal Education* (Baton Rouge: Louisiana State University Press, 1959).

General Discussion

Human biology is described as including molecular, cellular, organismal, and population studies. The proposed program in human biology at Dartmouth has not become operational for various reasons, including the inflexibility of the curriculum and the introduction of human biology into it at a time when students are already committed. Furthermore, the program is too long in duration. Another barrier is the difficulty of getting people to work together when there are anxieties over administrative and jurisdictional responsibilities.

The observation that medical schools are only interested in the "sick man" is incorrect. Medical schools have moved away from their overriding emphasis on the sick to a substantive approach to preventive medicine, and to improving the health of population groups as well as the individual. Examples cited of the preventive emphases included the role of pediatrics in well-baby care; prenatal and postnatal programs in obstetrics; and neighborhood health clinics operated by psychiatric departments. The failure of undergraduate schools to recognize this change emphasizes the serious lack of communication between them and the medical schools.

Man is not the exclusive unit for the study of all biological phenomena. Courses in cellular biology may be constructed effectively, based on drosophila, *E. Coli,* and other biological units.

A critical need exists to advance our knowledge of the functions of cells, tissues, and organs, as well as their interactions in health and disease. Such programs should be based in the medical schools to encourage the study of their relevance to health promotion and to intervention in critical disease problems.

Problems related to teaching human biology include the redundancy of the curricula; diversity in the backgrounds of students; teaching the basic versus clinical sciences; little acknowledgment of the importance of basic sciences in medicine; the students' lack of control over their educational programs; and the "premedical syndrome." The lack of appreciation of the importance of the basic sciences becomes more serious as the student progresses through medical school.

The "premedical syndrome" in students manifests itself in their concentrating their efforts so narrowly on gaining admission to medical school that major educational opportunities are lost; they choose non-challenging, "sitting duck" courses.

The mechanics of the first year of medical school present major scheduling problems. Most students prefer to move ahead with the group rather than to "place out," due in part to the students' competitiveness and to their apprehension over missing significant material. There is little advantage to having a student place out of a course unless it affords him a true elective opportunity.

The bottleneck to the production of more doctors lies in teaching the "organ" or "organismal" aspects of biology, where there is a dearth of qualified faculty; there is, on the other hand, a quite adequate supply of faculty to teach the cellular, molecular, and biochemical aspects of biology. Teachers of the organismal aspects are primarily in the clinical departments, and are largely unavailable because of their deep commitments to patient care, research, and clinical training programs.

The historic question was raised: Are those students who are the most accomplished in the basic medical sciences destined to become the best physicians?

The quality of training in biology may be eroding as the ratio of students to faculty continues to increase disproportionately—the so-called "premed glut." The primary evidence for this is in the laboratory experience: students are not acquiring the needed scientific methodology. Some consider it essential for premedical students to acquire an appreciation of scientific methodology through an actual research experience.

The premedical glut is probably going to be with us for at least a decade; projections from applicants taking the Medical College Admissions Test indicate that the trend is apt to continue into the 1980s.

The question of clarifying the career choices of premedical students versus courses in alternate health professions is a difficult one: at Northwestern approximately 450 of 1,100 incoming freshmen declare themselves as premedical, even though they may be pointed toward other careers.

The redundant nature of the educational program stretching from the early college years through the basic medical science period has been a factor in students seeking research careers; they have been turning away from medicine to enter graduate training in biology. Today, with the tight placement market for Ph.D.'s, these students are once again applying to medical school.

Early admission to medical school may result in the student letting down very perceptibly, as final-year medical students do after selections for internships have been completed. There may, however, be a different reaction if early acceptance allows a student a full year of work, as compared to an extra two, three, or four months.

Somatic Cell Biology
in the Medical School Curriculum *

THEODORE T. PUCK

During the past fifteen years there has been a revolutionary growth in studies of somatic mammalian cells that have application to a variety of branches of medicine. While aspects of cell structure and biochemical and physiological activity have traditionally been studied in departments of anatomy, biochemistry, physiology, pharmacology, and microbiology, it has now become possible to consider the study of mammalian cell biology as a coherent discipline, with particular applications to genetic disease, cancer, and differentiation, turnover, hormonal action, and analysis of the action of agents such as radiation and toxic chemicals.

In recent years the mammalian cell has been approached by means of the methods originally developed for microbial genetics. This has made possible genetic biochemical studies on somatic mammalian cells which have greatly enriched our understanding of this basic building block of the body. The following kinds of studies can now be carried out routinely on these cells, and each has contributed to the burgeoning field of mammalian cell biology and its medical application:

1. Establishment of long-term cultures from biopsies taken from an extremely wide variety of human tissues or those of other mammals. These can be cultivated in vitro for long periods, permitting many different kinds of studies.

2. Single-cell plating of somatic cells, such that each grows in isolation to produce a clonal colony. This permits recognition and isolation of clonal strains of mutants that can then be used for genetic experiments in which the mutated cells provide the markers to isolate particular kinds of cell strains and illuminate mechanisms of genetic behavior.

3. Quantitative measurement of survival after exposure of cell popu-

---

* Contribution #142 from the Eleanor Roosevelt Institute for Cancer Research.

lations to physical, chemical, and biological agents, by counting the number of colonies that developed when single cells are inoculated into a Petri dish. A "colony-forming" or "plating" efficiency can be determined as in microorganisms. This has proved to be a reliable quantitative measure of the reproductive capacity of cell populations under specified conditions. It has permitted accurate measurement of the effects of agents such as high-energy radiation, viruses, and drugs. These methodologies have resulted in the substitution of reproducible, quantitative measurement for uncertain qualitative observations on cell growth in vitro.

4. Determination of the chromosomal constitution. A new field of medicine, medical cytogenetics, has developed as a result of the availability of simple rapid means for the determination of chromosomal constitution. The chromosomes now not only can be counted and characterized by overall length and centromeric position, but can be identified by means of the new banding techniques. These permit even small deviations from the normal human karyotype to be recognized. By the combination of chromosomal delineation with the methods of cell growth in vitro described in paragraph no. 1, it becomes possible to diagnose the presence of chromosomal disease early in pregnancy, so that parents may if they wish terminate a defective pregnancy and initiate a new conception. These methodologies have given new powers to medicine, alleviating some of the terrible uncertainties connected with the possibility of giving birth to a seriously defective child. Diseases that can be approached by these methodologies include mongolism (Down's syndrome), Turner's syndrome, Klinefelter's syndrome, and the triple-X and XYY syndromes. Since these diseases are present in approximately 0.5 per cent of all human live births they constitute an extremely serious concern of medicine.

5. The nutritional requirements for growing mammalian cells *in vitro* can now be defined. Thus the growth characteristics of normal and malignant cells can be compared by means of their growth requirements. Moreover, comparison of growth requirements of single cells with those of large cell populations is possible, and furnishes a precise means for defining certain types of cell-cell interaction.

6. Single gene mutations can be produced in mammalian cells, and clones containing these genetic markers can be isolated and established as reliable cell stocks. The availability of such materials has made possible a wide variety of genetic and genetic-biochemical studies in mammalian cells in vitro.

7. By means of the cell hybridization technique developed through the individual contributions of the laboratories of Drs. Okada, Barski, Ephrussi, and Harris it has become possible to produce the equivalent of sexual crosses in somatic mammalian cells. Thus a whole gamut of genetic operations become possible with mammalian cells in vitro. These techniques greatly extend the power of genetic and molecular biological studies. Stable hybrids have been produced between cells of the same and different species, and indeed it has been possible to produce stable reproducing functional hybrids between cells of different biological classes such as the Chinese hamster and the chick. These techniques have therefore eliminated one of the traditional barriers for genetic analysis, since, in classical genetics, mating and genetic analysis can only be carried out between members of the same species. It is now possible to produce stable crosses from widely unrelated parental cells, thus making possible many new types of studies.

8. Genetic analysis of a wide variety of human disease conditions can now be undertaken. The chromosomal diseases have already been alluded to. An example of a single gene disease that has been illuminated by such means is xeroderma pigmentosum, a condition involving extreme sensitivity of such patients to ultraviolet light. This situation has been shown by in vitro studies to be caused by a gene defect that results in loss of enzymes needed to repair radiation damage.

9. These studies have illuminated conditions such as the mammalian radiation syndrome, which, before the advent of these methods, were difficult to understand. They have demonstrated that contrary to previous beliefs the mammalian cell is far more sensitive to reproductive death than are cells of microorganisms. The mean lethal dose of X-irradiation has been shown to be virtually constant for cells of different tissues of the mammal, and indeed to be very similar for different mammals. The reproductive death has been demonstrated to be due to chromosomal damage. These facts have permitted reinterpretation of the mammalian radiation syndrome so as to produce a simple explanation of many of the complex events attending whole-body and partial-body irradiation of mammals. Such considerations have made it possible to understand the magnitude of the doses needed to produce effective radiotherapy in tumors, and have aided in the understanding and management of cases of human irradiation injury.

10. By means of cell hybridization in vitro it becomes possible to determine relationships of dominance and recessiveness between allelic genes in mammalian systems rapidly and conveniently.

11. Through the use of interspecies cell hybridization, particularly between human and mouse and human and Chinese hamster combinations, it is possible to determine the human chromosomes on which particular genes are carried. This development is due to the fact that cells such as those of the mouse and Chinese hamster, when hybridized with a human cell, tend to lose unnecessary human chromosomes. If a Chinese hamster cell with an auxotrophic gene that gives it a special nutrilite requirement is hybridized with a human cell, and if the hybrid is grown in medium lacking this nutrilite, the human chromosome containing the needed gene will be retained in the clones that continue to grow. Identification of this chromosome then suffices to fix the chromosomal localization of the gene that was deficient in the original Chinese hamster cell.

12. Hybridization permits complementation analysis, which can determine whether two cells with the same phenotype have the same or a different genotype. For example, cells with the same nutritional deficiency are hybridized and then tested to see whether the hybrid does or does not display the nutritional deficiency. If it does, it may be presumed that both parental cells would be defective in the same gene. If the two cells complement each other, however, so that the hybrid no longer displays the given nutritional deficiency, the cells must be deficient in different genes, both of which can cause the same nutritional requirement. In this way complementation classes of cells displaying the same phenotype, but different genotypes, can be identified. Application of this procedure to cells of patients with genetic disease would appear to have great importance in diagnosis as well as in the biochemical understanding of the metabolic derangement.

13. Mutagenesis by applied agents can be quantitatively measured in somatic mammalian cells. This makes it possible to screen food additives, drugs, and environmental pollutants for their mutagenic activity, and to compare the degree of mutagenesis and carcinogenesis exhibited by the same agent.

14. It has become possible to search for and analyze genetic regulatory mechanisms in mammalian cells. It has been shown, for example, that one of the D-group human chromosomes carries a gene capable of activating a region of the Chinese hamster genome so as to cause the appearance of specific esterase activity previously not displayed by either parental cell. Similar examples of both positive and negative control of gene action exhibited by hybridization are being reported from several laboratories.

15. Precise methodologies have been developed for identifying specific cell-surface elements. Specific antibodies have been produced to such molecules, and their roles in specific cell-cell recognition, their chromosomal sites, their mutability, and their action in malignancy are under intense study. Exciting areas of study are opening up in the elucidation of the nature of the normal differentiation process and the abnormal patterns that constitute cancer.

16. The role of hormones on mammalian cells is being elucidated. The molecular pathways of action of the steroid hormones, on the one hand, and those involved in cyclic AMP dynamics, on the other, are under extensive examination. Explanations of the regulatory action on cell growth and the inducibility of specific macromolecular biosynthesis in particular cells is emerging. The elements controlling specific cell morphology are beginning to be elucidated. The application of these new developments to the understanding of pathological processes has barely begun.

17. Methods have been developed for measurement of cell turnover in specific tissues and isolation of the biochemical determinants that influence turnover in health and disease.

18. It has been possible to prepare mutants affecting different steps in a single biochemical pathway in mammalian cells. Such a series of mutants offers promise in elucidating the biochemical pathways of cells in different tissues; uncovering the mechanisms that regulate the rates of individual steps in those pathways; and helping to provide rapid identification of the site of specific blocks that are established under the action of trauma, drugs, or the presence of various genetic diseases.

*　*　*　*　*

This list could be continued. The point is amply demonstrated, however, that the revolution in mammalian cell biology currently underway promises a new revolution in medicine in which vast areas of empirical knowledge would appear to be on the verge of new interpretation and understanding. The powers of medicine will undoubtedly become enormously magnified. The teaching of science in the medical schools must prepare students for the new era that lies ahead.

General Discussion

GEORGE STREISINGER

The purpose of teaching the basic sciences is threefold: to acquaint students with specific subject matter; to review pertinent abstract data; and to have periodic critical tests of models. The problems include redundancy of curricula; diversity of backgrounds of entering students; lack of appreciation of the importance of the basic sciences; lack of control of a student over his progress; and the "premedical syndrome," which was described as "playing havoc" with undergraduate programs and eroding student values.

The desirability of medical schools eliminating from their curricula subject matter that students have mastered in their premedical programs was reviewed. This might lead to the elimination of one year or more from the medical school curriculum. It was pointed out, however, that in those schools whose students are allowed to "place out" of courses, this does not accomplish any shortening of the curriculum because of the variety of courses encompassed in the first year. The students who place out are offered more advanced courses in the same discipline or in alternate areas. The mechanics of the first year do not encourage acceleration. The use of qualifying examinations to determine when a student is ready to enter his clinical clerkship was proposed. Placing out may be of no value unless the student uses that time to pursue his own interests.

The "premedical syndrome" was cited by one participant as the major problem in trying to provide a good education to medical students and biology majors. With the numbers of well-qualified premedical students increasing steadily, there is an urgent need to establish real alternatives to medical school and relate them to premedical education.

The question was asked: What exactly is a premedical student? At a major private university, for example, 450 out of approximately 1,100 entering freshmen state that they are premed, even though they are pointed toward careers other than medicine.

The anecdote was told of a premedical student who requested an opportunity to enroll in a special undergraduate research course. When the professor asked about his purpose, he replied: "I would like you to get to know me better so that you can give me a strong recommendation for medical school."

It was suggested that the basic sciences constitute the bottleneck in the expanding system of medical education. More specifically, the cellular, molecular, and biochemical aspects can be taught to large numbers of students. There may, however, be insufficient faculty for teaching at this organismal and organ level of biology. Other participants pointed out that this requires cooperative efforts with faculty from the clinical departments.

The grafting of a new medical school onto an existing department of biology may raise problems akin to those found in the "two cultures." There are changing expectations; varying ideas about what constitutes a proper teaching load; different standards of living; and different salary levels.

The question of early admission to medical school is a difficult one. If it is made "provisional," very little is accomplished in terms of encouraging greater freedom of choice by the student. On the other hand, if it is a final commitment there is a major risk that a student may slack off in his academic efforts. A parallel situation was drawn in regard to the internship: once the decisions on internships are in the computer, students let down perceptibly.

Preprofessional and Professional Curricula in the Health Sciences: A Continuing Experience

PAUL A. MARKS

It is desirable at the outset to indicate the objectives of developing a more effective continuum between undergraduate and professional education—a matter in which there is an element of judgment for each faculty member. I am persuaded that a major objective should be to improve the quality of the educational experience in the biological and behavioral sciences and other areas relevant to the health sciences. The extremely rapid expansion in our understanding of life processes, the changing character of our social institutions, and the changing perceptions of what constitutes good health care make certain curricular considerations desirable, if not imperative. The scientific basis for the health sciences requires more, not less, rigorous basic science education. In professional education for the health sciences we cannot focus only on the biological sciences: the behavioral sciences, economics, mathematics, political science, philosophy, and other disciplines are also clearly important.

A second objective of curricular planning in continuum should be to provide for increased flexibility and for branching in the students' career objectives. We are going through a period where the number of applicants to medical school far exceeds the available openings. Well-qualified and highly motivated students are not being given opportunities to acquire a medical education. We have been very slow in recognizing that the potential for satisfying the drive of a student who seeks to serve in the health professions may exist in training him or her for a profession other than medicine. On the health sciences campus at Columbia, of the 1,400 full-time students only 580 are candidates for the M.D. degree.

A third objective should be to avoid unnecessary duplication in undergraduate and professional educational experiences. And, finally, a fourth should be to introduce into the undergraduate curriculum broader opportunities in human biology and related areas for all students.

A General vs. Professional Education

In considering a preprofessional or professional curriculum in the context of a continuing educational experience, it is essential that we emphasize the importance of general education. This is a complex problem that, stated in the briefest of terms, should aim "to liberate the powers of the individual by disciplining them; and that discipline in turn immediately relates him to values of his culture and certain social necessities." [1]

It should not be a primary objective of continuum curricular planning to achieve an overall reduction of the training period for a medical degree. It is likely, however, that better articulation between undergraduate and health science schools will, for some students, facilitate a shortening of the conventional four years of either undergraduate or professional training—or even both. In this connection it is interesting to note that the federal administration is rethinking its policies regarding programs that provide financial incentives to medical schools to shorten their curricula. In a recent speech Assistant Secretary of Health Charles Edwards stated: "I think that clearly we have moved beyond the point at which concerns about a shortage of M.D.'s were genuine if somewhat exaggerated." [2] I cannot help but comment that this position differs considerably from the one the administration took less than two years ago when it provided financial incentives to medical and dental schools to accelerate the production of more doctors. [3]

For several years a great deal of concern has been expressed about the relationship between general education and professional education. [4] More specifically, it has been suggested that the process of obtaining a general education is in some way separate from, and even inimical to, that of acquiring a professional education. This issue has been and may continue to be used to discourage efforts to establish better communication and coordination of undergraduate and professional educational experiences in the health sciences. General education and health sciences education are not mutually exclusive; in a very real sense they are quite interdependent. Medicine and the related professions are social enterprises: I need not dwell on this point. Ideally, general education must be a continuing element in the health sciences curriculum; realistically, however, this is hard to achieve in a structured fashion. We must provide opportunities for general education throughout the undergraduate curriculum and introduce it more effectively into our professional health sciences schools.

A Continuum in Curricular Planning

Let me turn to specific considerations for a continuum in curricular planning for undergraduate and graduate health sciences education. What follows is in part a result of discussions among a group that has included Arnold Relman of tht University of Pennsylvania; Richard Rifkind of Columbia; Thomas Roos of Dartmouth; Leon Weiss of Johns Hopkins; and Alan Kropf of Amherst. I, however, must take responsibility for the material as presented.

Several difficulties are encountered in considering a continuum in curricular planning. I shall not attempt to catalog all the issues that may have to be resolved for they vary from one institution to another. They will be different for a university with undergraduate, graduate, and professional schools; a university with undergraduate and graduate schools but no professional schools; and an undergraduate or professional school with no university affiliation. The problems include the relatively rigid curricular structure of the first year of professional school; the sharp division between undergraduate and medical school experiences; the need to coordinate the calendar; admissions policies; and, possibly, faculty assignments and compensation. While I have no easy answers for coping with these questions, I do think we must first attempt to define our curricular goals and then deal with the constraints of the interface beween undergraduate and professional schools.

A Sequence of Courses

In this context we can address the issue of a desirable sequence of courses in the sciences and related areas that could be recommended for students who wish to pursue careers in the biological and health sciences. These course requirements can be considered without regard for the fact that a break does exist between undergraduate and professional school. One possible sequence of courses in the sciences is illustrated in Figure 1. It includes certain courses traditionally considered desirable prerequisites for health science professional schools, such as general biology, general and organic chemistry, physics, calculus, and English. More advanced courses given at a university-wide level might include cell biology—which would cover histology—molecular genetics, biochemistry, behavioral sciences, and systemic biology—which would cover anatomy, developmental biology, and physiology. These courses are essential for dissection anatomy, pharmacology, pathology, and immunology. Such a sequence of courses may have to be provided at two levels: one designed for M.D.,

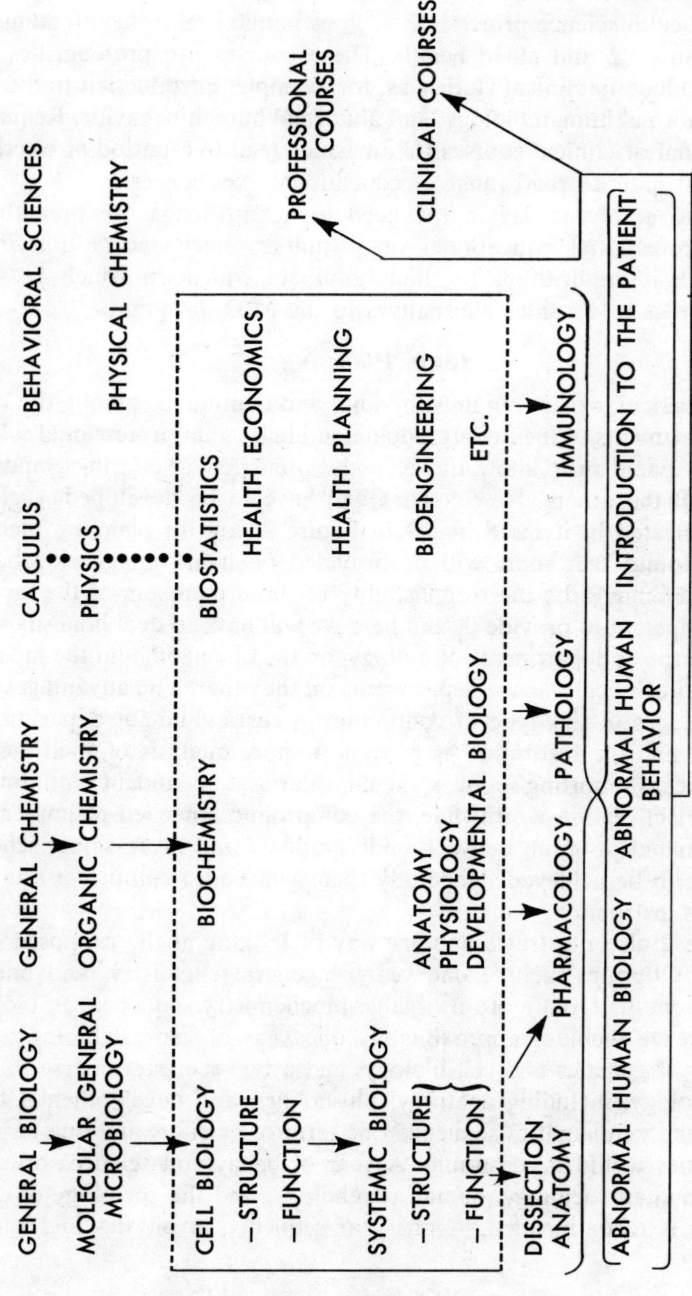

Figure 1. One possible sequence of courses in the sciences

D.D.S., and Ph.D. candidates in human biology; the other for candidates in other health science professions such as public health, health administration, nursing, and allied health. These courses are prerequisites for such introductory clinical studies as, for example, introduction to the patient, abnormal human biology, and abnormal human behavior. Required professional or clinical courses follow which lead to a period of electives that could span a broad range of educational experiences.

I would emphasize again the need to acknowledge the breadth of health professional educational opportunities. Such recognition must carry with it implications for undergraduate education which provide opportunities to consider alternatives to the M.D. program.

Joint Planning

This curricular planning must be joint and continuous among the relevant departments of the undergraduate, graduate, and professional school faculties. Each must know the objectives, curricular offerings, and resources of the others. New courses will have to be developed, such as those indicated in items 8 and 9 in Figure 2, and in planning them it may be found that some will be provided on a university-wide basis. Instruction should be the responsibility of those members of the faculty most competent to provide it, and here we will have to deal honestly with the concerns of departments of biology, on the one hand, and the medical schools' preclinical science departments on the other. The advantages and disadvantages of this type of continuum in curriculum for departments, students, and the institution as a whole require analysis of such factors as facilities, supporting services, faculty salaries, and student recruitment. If the objectives are worthwhile, the compromises needed to implement a coordinated program between undergraduate and professional schools will have to be achieved. Hopefully there will be no compromise in the quality of instruction.

Figure 2 also illustrates another way of looking at the proposed curriculum. Offerings include one year of general chemistry, and one to one-and-one-half years of an organic-biochemistry sequence. In biology the sequence includes approximately one year of general biology; one semester of genetics and cell biology; and a two-semester course in systemic biology, including anatomy, physiology, and developmental biology. In the mathematics sequence one term of calculus and one term of biostatistics would be desirable. A year of behavioral sciences oriented toward human sociology, human psychology, and the sociology of communities is recommended. A one-year sequence in physics and a one-

1.	Calculus	General Chemistry	General Biology	English
2.	Physics	General Chemistry	General Biology	English
3.	Physics	Organic Chemistry	Genetics, Molecular	Behavioral Sciences
4.	x	x	x	Behavioral Sciences
5.	x	x	x	x
6.	x	x	x	x
7.	x	x	x	x
8.	Biostatistics	Biochemistry	Systemic Human Biology	x
9.	x	Cellular Biology	Systemic Human Biology	
10.	x	x	Dissection Anatomy	
11.	Immunology	Pharmacology	Pathology	
12.	Abnormal Human Behavior	Abnormal Human Biology	Introduction to the Patient	
13.				
14.				
15.		Clinical Years		
16.				

Figure 2. A proposed curriculum

Note: x = General education courses. The courses are listed vertically in sequence only to indicate probable prerequisites; the courses are not necessarily to be taken in a given semester, and horizontal relationships are not implied in this listing.

year sequence in English are also indicated. In addition, of a total of twenty-four- or thirty-two-term courses, that is, a three- or four-year undergraduate curriculum, at least seven to fifteen terms of general education courses are available.

In considering this curriculum it is important to recognize that the sequence in which students take the courses will be determined by their prerequisites. The university-wide courses such as biochemistry, systemic biology, cellular biology, and biostatistics are considered as offerings that are likely to be taken by students in the equivalent of the first year of medical school or the junior or senior year of undergraduate school. In considering the design of new courses we must have a sound data base with regard to existing curricular offerings. From this base we can move to define the objectives of specific offerings, keeping in mind the overall framework toward which we are moving, namely, an operational plan for the curriculum for undergraduate and health science schools. Implementation of this program clearly presents many problems, but I doubt whether our consideration of them need become so complex as to preclude our achieving our objectives. I do know that up until now communication among the undergraduate and basic science and clinical departments of the health science schools has not been adequate enough to permit us to provide the answers to these questions.

NOTES

1. W. T. deBary, *Seminar Reports: General Education and the Humanities,* vol. 1 (New York: Columbia University, 22 October 1973): p. 1.

2. Charles C. Edwards, "A Candid Look at Health Manpower Problems," *Journal of Medical Education* 49, no. 1 (January 1974): 19–26.

3. U.S. Government Comprehensive Health Manpower Training Act of 1971 (P.L. 92–157).

4. Daniel Bell, *The Reforming of General Education* (New York: Columbia University Press, 1966).

General Discussion

Too many students are totally committed to the belief that they will learn only if they take courses with appropriate university numbers. Students need to be encouraged to self-learning by reading on their own.

Representatives of about forty college biology departments have been meeting for the past several years to prepare summaries of courses in biochemistry, cell biology, genetics, and microbiology.

Teaching Biochemistry in the Medical Education Continuum: The Roles of Undergraduate and Medical Colleges

THOMAS M. DEVLIN

The education and training of future physicians is a continuum. For those who early identify medicine as a career goal it begins during secondary school, for others, not until late in the undergraduate college period. Many forces, including the explosion of new biomedical information and the need for more physicians, have obligated educators to define more clearly the objectives of the formal components of the educational continuum in order to reduce unnecessary redundancy and improve the effectiveness of the process.

It is appropriate to consider the roles and responsibilities of undergraduate and medical colleges in the teaching of biochemistry, for this discipline more than any other overlaps into all of the medical sciences. A review of the curriculum content for premedical and medical students leads to an appreciation of the variety of settings in which aspects of biochemistry are presented, ranging from undergraduate first-year courses in biology to the instruction of young physicians at the patient's bedside. Because of its involvement in such a diversity of other specialities, one might even question whether biochemistry need be taught as a separate disciplinary course. A careful scrutiny of the depth of biochemical content and emphasis in these programs, in comparison with the breadth of biochemistry, leaves little doubt about the requirement for future physicians to participate in a formal course as a foundation for future development. Biochemical subjects taught in the context of other disciplines are frequently presented in a superficial and generalized manner, limited in scope to those aspects pertinent to a particular discipline. This is an unsatisfactory mechanism for acquiring a biochemical knowledge base, and leaves the student with a disjointed and incomplete foundation.

The question therefore is: Where in the medical education continuum should students be challenged by the formal rigors of a course or courses in biochemistry? This is not an easily resolvable problem, since both undergraduate colleges and medical schools have a responsibility in biochemical education. The complexity of the system for training a variety of students does not, and should not, permit a specific limiting role to one component, universally applicable to all educational and geographic situations.

The discussion that follows is restricted to a consideration of the formal presentation of a program in biochemistry. It should not, however, be construed as implying that biochemical subjects should not be presented in a variety of settings, such as is currently the case.

Students, Physicians, and Biochemistry

Before attempting to define the role of the preprofessional and professional aspects of the educational continuum, there are three concerns to be considered: (1) the student who is the recipient of the educational process; (2) the needs of the physician who is the product; and (3) the curriculum content, in this instance, biochemistry.

Students enter the medical education continuum at a variety of points and begin their professional education with different backgrounds. Students entering medical schools in the United States are *not* homogenous with respect to prior educational attainments: many have undergone extensive scientific training, whereas others enter with little thought of pursuing medicine as a career; attracted by the idea of developing intellectual self-confidence and gaining a better insight into the potentials of a career in medicine, the students elect this new direction midway through the educational continuum.

During the 1960s medical school admissions committees made a healthy move away from selecting only students who had completed traditional premedical curricula toward choosing students with backgrounds in the humanities and only minimal scientific experience. The rationale was the desire to train physicians with a wider spectrum of interests to fill new and ever-expanding roles in the health care delivery system.

A diversity of undergraduate colleges are involved in premedical education. The student's selection of a college is based on a variety of factors: economic, geographic, and sociological, as well as academic. For those who have decided on a medical career, the track record of the college or university in training premedical students is a consideration, but for many others it is not. Over 750 undergraduate colleges and universities are involved in preparing students for medical school. Studies conducted by the Association of American Medical Colleges on the entering medical school class of 1964 [1]—the latest data available—indicated that approximately twenty-five colleges supplied 30 per cent, and 155 colleges 75 per cent of the medical student body; 600 other colleges educated the remaining students. It would be untenable to propose that a few selected schools be designated as the only institutions that could train premedical students, even though this might be the most economical procedure. The health and future development of medicine depends on individuals with different backgrounds, interests, and experiences entering the profession.

One of the most difficult tasks in developing a curriculum is to define the educational and training skills that will be needed by the physician of the future. Obviously the curriculum cannot contain an in-depth experience in all of the potential topics that physicians may require; it must, however, be constructed on the basis of developing a foundation upon which the student can continue his or her education. In addition, it is necessary to train the student in certain areas, for without these specifics it would be difficult to advance into new fields. In the past, instructors in the preclinical science departments frequently selected material for presentation that they considered most fashionable or interesting, giving little thought to how the student would utilize the knowledge in the future. The pressure of time has now forced faculties into giving greater consideration to the needs of the student. In developing a biochemical curriculum the faculty must carefully consider the potential behavioral changes of the student, and how, when, and where he will utilize the information as a physician. It is not sufficient to merely present a "good" course in biochemistry; the content must recognize the special needs of the student. Determination of objectives does not preclude the education of physicians as responsible members of society able to cope with the changing science milieu both in and out of the profession, nor does it restrict the curriculum to training students in the specific clinical skills that physicians are required to have.

As in all of the biomedical sciences, the amount and diversity of knowledge in biochemistry that has accumulated over the last few decades is overwhelming. One need only survey the variety of biochemical texts and journals available, to read the biochemical discussions in the literature, or to participate in medical grand rounds to be impressed with the impact biochemistry has had on the biomedical sciences. Twenty years ago very few colleges offered even a single course in biochemistry; today many offer an undergraduate major consisting of a series of courses of varying degrees of specialization. Even at the secondary school level, biochemistry is a significant portion of many biology programs. Biochemistry cannot be presented in a single course; it should be taught at various educational layers. Thus one should not consider whether a single course in biochemistry is required, but, rather, how multiple exposures to courses of varying content and depth should complement one another in the medical education continuum.

Undergraduate and Professional Training Phases for Students in Nonintegrated Programs

The vast majority of students entering medical school now and in the foreseeable future will not have participated in an integrated or accelerated premedical program. Most will have completed a somewhat standardized curriculum, having taken an undergraduate course in biochemistry ranging from a two-hour-a-week survey to an in-depth course for majors. There is a wide variation of depth and content in these programs. If one states the specific biochemical facts required by the physician, even though in many cases inadequately derived, most undergraduate programs do not deal with many areas of physiological or clinical chemistry considered important by the medical school.

Over the last four years, for placement purposes we have evaluated incoming freshmen at Hahnemann in terms of their background and experience in biochemistry. During this period approximately 50 per cent of the students have indicated satisfactory completion of a course in biochemistry, cell physiology, or cell biology, but upon a challenge examination less than 5 per cent have had sufficient background to be exempted from the medical school course. The examination covers the usual major areas presented in a general biochemistry course, and does not cover exclusively those areas of medical biochemistry emphasized in our particular core program. These observations are similar to those of a number of medical schools.

An informal survey suggests that 50 per cent of incoming medical students have had some chemistry training, and that, surprisingly, over the last few years this figure seems to have remained steady or even to be slightly on the decline, whereas one might have predicted an increase. After completion of our course, students indicate that the emphasis and requirements of the medical school are so markedly different from those of undergraduate school that it would have been a mistake not to have taken the course. These observations do not necessarily signify that undergraduate courses are not equal to those of the medical schools, but that there is a variation in content and emphasis, as well as in student retention of knowledge.

It is not advisable, even if feasible, to systematize the educational process at all the feeder colleges in order to provide a uniform product. That would detract from the important function of the undergraduate school of putting students in contact with the broader aspects of the science. Why students do not do better in medical school placement examinations is difficult to determine. Obviously it could be the examinations, biased as they are toward a particular department, or lack of student preparation, but I believe the reason is more complex. While students may have participated in well-organized undergraduate courses presented by excellent instructors, they may have worked only for grades and not for long-term retention of content.

For those students who enter medical school with adequate backgrounds, it is the responsibility of the faculty to make available meaningful alternatives either in biochemistry or other disciplines. Students who have had no exposure to a formal course will continue to need a presentation in the professional training phase.

Thus the problem of redundancy is not as severe as some have suggested, and a large percentage of the entering medical student population will continue to require a course in biochemistry, either basic or in selected subjects.

Biochemistry as a Prerequisite for Admission to Medical School

One frequently heard proposal for shortening the medical curriculum is to make biochemistry a prerequisite for admission. A course prepared and presented to meet the behavioral objectives of future physicians would be quite acceptable, but one developed around the interests of a few faculty members would be inadequate and not prepare the student for his subsequent educational experiences. Undergraduate courses must fulfill the educational needs of students with a variety of career objectives; to bias these courses exclusively toward the needs of future medical students would be to abrogate an important responsibility.

With the already overbearing pressures placed on premedical students, and the resulting so-called "premedical syndrome," it is questionable whether biochemistry should be universally adopted as a prerequisite. It might, however, supplant organic chemistry as the undergraduate course frequently considered to separate out the acceptable from the unacceptable candidates for recommendation to medical school. Many medical students' woefully inadequate knowledge of organic chemistry indicates that their primary concern with that subject is to secure the best grade possible rather than to develop a lasting knowledge of it.

Many medical educators endorse exposure to biochemistry at the undergraduate level because it may develop in the student an appreciation of biochemical research, its methods, and direction—an objective that the medical curriculum has abandoned due to the pressure of time. Moreover, if most students entered medical school with adequate backgrounds in biochemistry, departments could initiate programs at more advanced levels.

Many schools already offer a two-phase biochemistry sequence: general biochemistry and medical biochemistry—the latter including subjects not presented in most college courses. Students with satisfactory backgrounds need not participate in the first phase. This approach is causing a change in emphasis in the teaching activities of many medical school departments of biochemistry, in that faculty members have to take a greater interest in the relationship of biochemistry to medicine and not satisfy their teaching responsibilities by merely stating fundamental concepts. In order to emphasize the relevancy of biochemistry to medicine, many medical school faculties now present clinical correlations and sequences based on diseases or organ systems [2, 3] rather than biochemical material alone. Development of visual aids [4] and self-study curricula [5] by biochemists in the medical environment demonstrates that they are responding to the changing backgrounds of students.

Biochemistry in Accelerated Programs

A variety of programs have been developed to shorten premedical and medical training by combining some of the educational activities in the two phases. Students in five- and six-year programs, such as those at Jefferson-Penn State,[6] Northwestern,* and Hahnemann-Wilkes,[7] are selected on the basis of their high level of academic achievement and motivation. The premedical program is often abridged to about two years, with students taking the standard four-year medical curriculum, including biochemistry, and competing with others from traditional four-year baccalaureate programs. In these cases, the teaching of biochemistry is better left to the medical school phase in order to permit the student more flexibility in the premedical curriculum.

The development of more integrative human biology programs, such

---

* See Arthur Veis's presentation in this volume (page 113).

as that between MIT and Harvard, has opened up new approaches whereby students have the opportunity to complete some of the preclinical sciences, particularly biochemistry, prior to entering the medical center. These programs require extensive interaction of the premedical and medical faculties in curriculum planning, and can be accomplished successfully only where cooperation is active and the common educational goals are clearly defined. This type of innovation should be encouraged as one option available to some students; in such cases biochemistry can most effectively be presented during the first two years in the context of a human biology major dealing with the normal human being.

The Biochemistry Department Based on the University Campus vs. the Medical Center

The growth of biochemistry has resulted in the development of a number of centers on university campuses where active biochemical teaching and research is conducted. One often finds such groups not only in the medical school but in departments of biology and chemistry, as well as in independent departments of biochemistry. Why then could not a university-based department assume the responsibility for all formal biochemistry teaching, including that required of medical students? The determinant factor in resolving this question should be the local situation. In a highly integrated university with excellent communication between the college of arts and sciences and the medical center, there is no reason why such a department could not be directly responsible for the training of the future physician; examples of this approach can be found at Michigan State and North Carolina.

A frequently observed difference between university and medical school science departments is the extent of departmental influence on curriculum content. In the undergraduate setting the selection of topics for emphasis is often based on the judgment of a single faculty member, even though the general course content is decided by the faculty, whereas in most medical schools the faculty of the department reaches a consensus concerning the importance of specific subjects, with special reference to the needs of the students. In addition, many medical schools have strong faculty curriculum committees that serve to prod the departments into fulfilling the specific requirements of the future physician.

Since many university medical schools are geographically separated from the parent institution, this serves to create two different and distinct academic environments. An extensive educational or geographic separation weakens the influence of the medical faculty on the determination of the objectives of a biochemistry course. Some medical schools are free-standing units that do not have the resources of university science departments, and in these cases the medical faculty has little influence on the content of biochemistry courses presented on campuses not directly affiliated with the medical school. These problems are greatly reduced in more integrated programs, where the faculties of the college of arts and sciences and the medical school cooperate in the development of the curriculum.

One disadvantage of biochemistry being presented outside the medical center is the inhibition this has on the development of interdisciplinary approaches in preclinical training, such as the Case Western Reserve model. It is unlikely that university-based departments will be willing to relinquish control over content and format as required in these programs. The presentation of biochemistry in the medical school, along with the other basic medical science departments, also encourages the delegation of parts of the basic medical science curriculum to departments on the basis of their individual strengths. At Hahnemann, our Department of Biochemistry assumes responsibility for teaching aspects of gas transport and pH regulation, subjects that are frequently the province of physiology departments.

Conclusions

I have attempted to indicate the important roles of both undergraduate colleges and medical schools in the teaching of biochemistry in the continuum of medical education. Even though this discussion has concentrated on the presentation of formal courses, there are, and I hope there always will be, a variety of settings in which biochemistry will be taught to future physicians, including formal undergraduate and medical college courses, and presentations in other disciplines, as well as in the clinical setting. There is not a single correct or necessarily best approach to the education of physicians; each must be measured against its objectives, the diversity of the students' backgrounds, and the academic environment.

Undergraduate colleges should continue to develop superior biochemistry courses, but they should not be directed exclusively toward potential medical students. Such courses should not attempt to replace those in the professional phase, but should complement the more human physiological approach presented in the medical school setting.

Medical school departments of biochemistry will continue to play an important role in preparing students who come to medical school inadequately prepared in biochemistry, and in presenting formal courses on topics directly related to the needs of medical students. Biochemical faculties in medical schools should also serve to complement the clinician in the teaching of the basic medical sciences in specific medical areas. Basic medical science departments willing to accept the responsibility to participate in the total education of the future physician, in contrast to being concerned only with the presentation of a single first-year course, should have no difficulty in justifying the need for and role of such departments in the medical school.

Where appropriate, university-based departments of biochemistry can take a major responsibility for the teaching of medical students, but only when the department is willing to accept the input of the medical school in setting course objectives; when the department will actively function in planning the medical curriculum; and when faculty members participate in the activities of the medical center.

In the entire continuum of medical education both undergraduate and medical school faculties must continue to seek ways to interdigitate, complement, and make their program more flexible so that medical education will be able to meet the increasing demands of the future.

NOTES

1. Datagram: "Undergraduate Colleges in Relation to Medical School Applicants and Enrollment," *Journal of Medical Education* 42, no. 83 (1967).

2. D. Rubinstein, "Stimulation of Positive Attitudes on Biochemistry by Incorporation of Clinical Presentations," *Journal of Medical Education* 47, no. 3 (1972): 198–202.

3. M. Saffran " 'Relevance' in the Medical Biochemistry Course," *Journal of Medical Education* 46, no. 12 (1971): 1080–86.

4. H. B. White, T. M. Smith, and L. L. Sulya, "Self-Instructional and Audiovisual Method of Teaching Biochemistry Laboratory," *Journal of Medical Education* 48, no. 10 (1973): 939–44.

5. R. A. Weismann and D. M. Shapiro, "Personalized System of Instruction (Keller Method) for Medical School Biochemistry," *Journal of Medical Education* 48, no. 10 (1973): 934–38.

6. P. A. Herbut et al, "The Jefferson-Penn State Accelerated Medical Student Program," *Journal of Medical Education* 44, no. 12 (1969): 1132–38.

7. W. W. Oak et al, "Hahnemann-Wilkes Program: Six-Year Plan to Train Family Physicians," *Pennsylvania Medicine* 76, no. 3 (1973): 40–41.

General Discussion

The value of pretesting was emphasized; participants were urged to evaluate the students' knowledge at the beginning rather than at the end of a course. This would permit a revision of the curriculum to match student achievement, and identify students who need tutoring. Thus pretesting should not be considered as useful only for determining advanced placement.

Over fifty medical schools accept results on the biochemistry advanced placement tests of the Association of American Medical Colleges. Other placement tests include those of the National Board of Medical Examiners and the American Chemical Society, as well as departmental examinations. A placement examination should be based on material that the students are expected to know at the end of the course. In biochemistry it should include fundamental chemistry so that those requiring remedial work can be identified. A placement examination may also improve the attitude of students toward a course, because they can identify their deficiencies and the new knowledge they need to acquire.

The percentage of first-year medical students who place out on biochemistry pretests varies enormously from one medical school to another —from a handful, to one-third, to as high as 70 per cent of the students. There is a need to identify a basic core of information the students must master.

The varying levels of student achievement may be met by replacing a single large course taught by committees with a cluster of four or five courses graded to match students' needs and backgrounds.

There is no unanimity of opinion in the medical schools as to an ideal syllabus for a course in biochemistry.

The importance of a strong background in human anatomy as a prerequisite to human physiology courses was emphasized.

The Keller method of independent self-study, with a series of units of self-learning and self-evaluation, has been used by several departments of biochemistry (see note 5 on page 67). It is preferable to have a uniform time schedule for completing the self-study units, otherwise students may progress at totally different speeds.

Processes in Development and Evaluation
of a Curriculum, 1629 and Beyond

THOMAS HALE HAM

Unfortunately, the curriculum cannot, like Pikes Peak, remain resolute in stone forever; rather it is a continuing journey. To arrive at the phenomenon we call curriculum requires about seven steps in strategy, two in tactics, and at least three in evaluation. These twelve steps are processes used in exploring the medical student, which are comparable to those used in exploring the biological sciences and clinical medicine. These processes allow.the problem approach, with the benefits that come from inquiry; from the development of criteria for understanding cause and effect; from methods of measurement and of recording; from establishing and testing hypotheses; and from analysis and conclusions. In other words, the beginnings of the scientific method may be applied to learning medicine. But none of it can be hurried, and most of the processes must begin by simple description and classification as disciplines.

The processes to be considered here include approaches to the needs of medicine, of the learner, and of the faculty, and to the development and evaluation of curriculum. It is a humble beginning because our methods are so limited. We shall consider the learner as the client as we develop a strategy and tactics of the educational environment, the program of education, and evaluation.

Changing Needs of Medicine

Certainly the need of the patient is the integrating force at all times, and we recognize patient care as primary, preventive, and specialized. Multiple health agencies have arisen from original hospitals; delivery of health care has become local, regional, and governmental in orientation; and costs are key issues, as are systems for providing care. Out of the many facets of health care systems comes the one final, common path for medical education: the *individual patient* and his or her needs for health care. We begin here; we always return here.

A hierarchy of needs is subtended from the patient, continue to grow in number and in complexity, but ultimately relate back to patient care, so that none is mutually exclusive of the other although it often appears so. The specialty hospital, for example, has grown up in a century to furnish high degrees of expertise in more than twenty fields of specialization and to educate many levels of students. Much of the system is episodic and based on fee for service. Yet the task of medicine has changed markedly since the introduction of antibiotics.[1] In the past, the first six out of ten deaths were caused by infections or infection-related disorders; today only one out of ten deaths is due to infection, mainly in the newborn. Chronic illnesses now constitute eight of the major causes of death, and thus a whole new system of care and of training in prolonged patient care is required. Patient care has become a major challenge for all concerned, including the medical educator as a colleague in planning.

Education in the health professions now includes the training of the medical student and students of allied health professions; learning by practitioners of the health professions, by multiple specialties, by community group, and by the many fields that require qualifying examinations for practice. More and more the university is developing health science education in cooperation with specialized licensing groups. Education and evaluation of performance in patient care are becoming major responsibilities, complementary to each other, and demanding continuous planning.

Roles of Biological Sciences, Clinical Sciences, and Patient Care

Obvious needs now include the science of medicine, namely, biomedical research and clinical research, which are the foundations for the future understanding of disease and treatment, and systems of working with patients. Faculty members must work in a productive environment in order to make possible the biological and clinical advances we must have. Add to this the needs for research in the delivery of medical care, including the study of costs, and we have traversed a full circle back to the patient. While much but not all of this research will be university-oriented, the universities will be vital members. The changing requirements of departments to meet multiple demands require planning and recognition in order that the foregoing needs can form a compatible hierarchy. Previously, one particular need for medicine appeared to be mutually exclusive of others, but now we recognize all of them as a continuum with a beginning, a sequence, and an interdependence.

The history of medicine allows us to understand the current period and to project into the future (Table 1). For example, primary care and medical education in the United States began in 1629 in Salem, where the first apprentice served under a medical preceptor for three years in order to qualify for practice. The requirements for primary care and preceptorships are with us today. John Morgan in 1765 enunciated the need for medical education in the colonies, and was immediately invited to establish the medical department of the College of Philadelphia, the first medical school in the United States.

In the 1800s there was an enormous increase in the number of proprietary medical schools. The medical department of Johns Hopkins University, founded by William H. Welch in 1893, and Abraham Flexner's Report of 1910 [2] signalled the beginning of departmental growth and research and of sequential university education of the student. Thirty-five years after the revolution created by the Hopkins-Flexner advances, in 1946 the needs of the student were being rediscovered and reevaluated by several schools of medicine. More recently the need for the delivery of health care to the patient and to the community have been rediscovered, as will be discussed later. We have come full circle from 1629 in Salem, in three-and-one-half centuries, to face a series of questions that require answers, just as they did then. These questions lead inevitably to the development of a plan that will relate the learner to medical education, based on a history that allows us in part to predict the future.

TABLE 1. HISTORY OF NEEDS

1629	Medical preceptor, community at Salem; patient care
1765	University of Pennsylvania—didactics, demonstration, anatomy
1800s	Proprietary schools; universities
1893	Johns Hopkins University
1910	Flexner Report; needs of departments—biological sciences, clinical sciences
1946	Needs of the student
Current Period	Needs of the patient; needs of the community

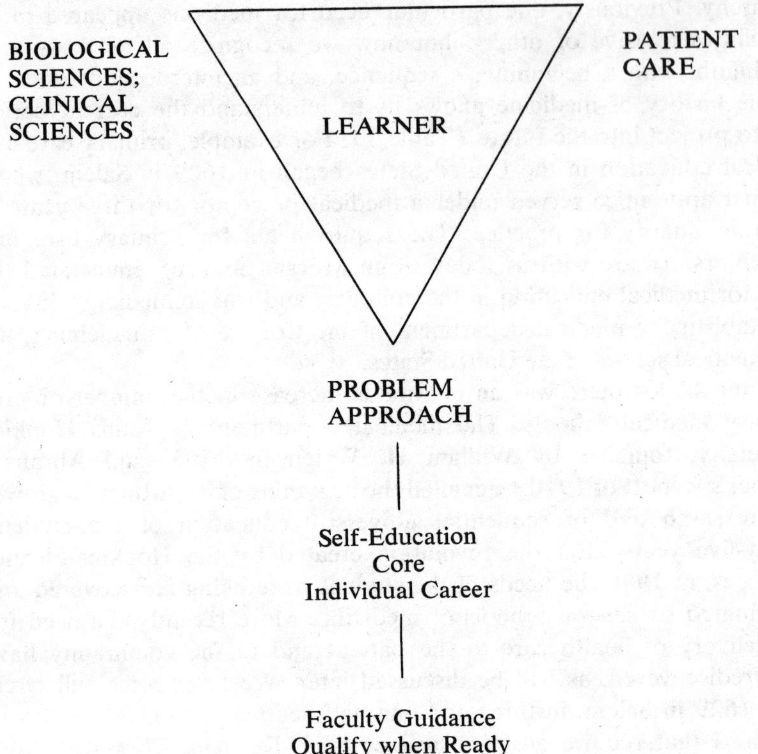

Figure 1. An overall plan to relate the components of medicine to the learner and the guidance of the learner.

Development of an Overall Plan

The overall plan of medical education must begin with the learner and indicate the essential needs of the biological sciences, clinical sciences, and patient care. The student is shown as being guided by the problem approach, with self-education in the core material and the individual career (Figure 1). The faculty provides guidance and qualifies the learner when ready. (This is a model for planning.) The learner is the client of the medical school and study plans are developed for him by the guidance systems so well described by Ralph W. Tyler.[3] These are based on objectives, learning experiences, and evaluation (Figure 2). Discussion of these processes will be expanded later.

There is good evidence from student and faculty responses that the

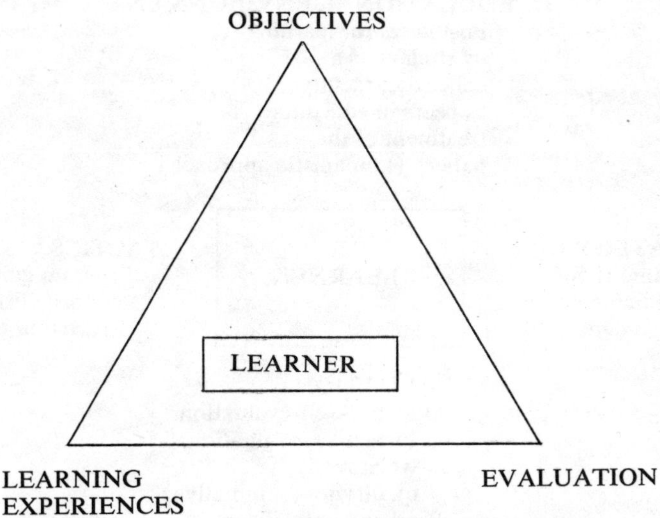

Figure 2. Development of guidance systems.
Source: Ralph W. Tyler, "Basic Principles of Curriculum and Instruction,"
Syllabus for Education 360 (University of Chicago, January 1950).

development of an overall plan requires a series of essential processes that can be identified, understood, and accepted by all participants, used for orientation in discussions, and for projections of where one's plans are leading. A brief outline indicates a sequence of four processes essential to the development of a program of education (Figure 3). The curriculum thus becomes a tactical translation of the preceding steps.

Faculty Participation in Developing a Program

In 1946 Joseph T. Wearn convinced the new department heads of the School of Medicine of Western Reserve University to delegate three responsibilities to the General Faculty: (1) instruction; (2) student affairs; and (3) interdepartmental cooperation. Considerable discussion resulted in the agreement in 1950 that the Committee on Medical Education would carry out the following functions:

1. EDUCATIONAL ENVIRONMENT
Postulates for learning
by student as a
colleague; faculty
as peers in education;
treatment of the
patient (humanistic approach)

2. STRATEGY
Planning for
medicine needs
of student

LEARNER

3. TACTICS
Program guidance
systems; curriculum
description

4. EVALUATION
Student—self-evaluation,
formative, evaluation
when ready,
qualifying (summative)
Program—effectiveness,
significance,
limitations,
student responses

Figure 3. Sequential steps in the development of a program of education.

Article XIII. Program of Medical Education

Section 1. *Approval and Conduct of a Program.* It is believed that the program of medical education can be evolved in a continuing and experimental manner with freedom of discussion, with opportunity to disagree and with opportunity for departments, faculty and students to cooperate.

It is understood that the Committee will make detailed recommendations concerning the program of medical education for approval by the General Faculty. It is understood that the Committee will have no direct administrative authority to change or enforce a program of medical education. When approved by the General Faculty, the program will be conducted by members of the Departments and not by the Committee which would continue, however, to evaluate the program.

Section 2. *Details of a Program.* When a program is reported to the General Faculty, it is understood that certain features will be considered and reported in detail, such as the following: (i) object of the program, (ii) methods for conduct of the program, including the delegation of authority when more than one department is involved, (iii) detailed description of the program with a schedule of hours required for the conduct of the program, (iv) if experimental trial on a small scale is necessary, a report of such a trial will be included, (v) methods of teaching, (vi) method for evaluation of the program, (vii) estimated time and cost of the program.

Over the period 1950–52, more than 300 members of the General Faculty developed educational objectives and prepared detailed plans of the organizational structure that is still in existence today (Figure 4).[4] There were unlimited opportunities to plan, to discuss the implications for each department, to develop models of interdepartmental teaching, and to consult experts, students, and members of other disciplines. These were democratic and investigative approaches to an enormously complex subject, but communication and evaluation were everpresent. Faculty participation was extraordinary.

After twenty months of intensive planning and five years spent in developing an overall plan, agreement was reached on a revised program of medical education.[5] The "Objectives Concerning Curriculum"[6] formed a strategic concept that has acted as a guide for twenty-three years.

Students and the Educational Environment

The learner, as stated earlier, is the client of the university. The most powerful unifying force of the university effort is the problem approach to the education of one person—the student. The faculty become the strategic designers of appropriate exercises involving real issues that are relevant to the student's need for learning, bring multiple disciplines together to define objectives, evolve methods of approach, and develop criteria of evaluation. On the student's part, the problem approach authorizes him to take active responsibility for studying problems and learning by participation in the process of search, analysis, and reporting, thus making the faculty and the learner-client colleagues in learning.

Combined with the problem approach is the faculty serving as coaches in the education of the learner-client. Emphasis is on the student's self-education and self-evaluation with faculty guidance as the student seeks answers and learns. *There is enormous power for learning by the student, based on his initiative and on guided inquiry.*

It is essential to "discover" the student as a colleague, as this allows the teacher to serve in the exciting roles of coach, friend, ombudsman, colleague, helper, critic—all these function for the student's needs in medicine as a biological science and for his future needs as a physician in caring for his patients.

Thus the teacher is providing guidance for the student, who is given authority to move out in his learning. This strategy reverses the usual

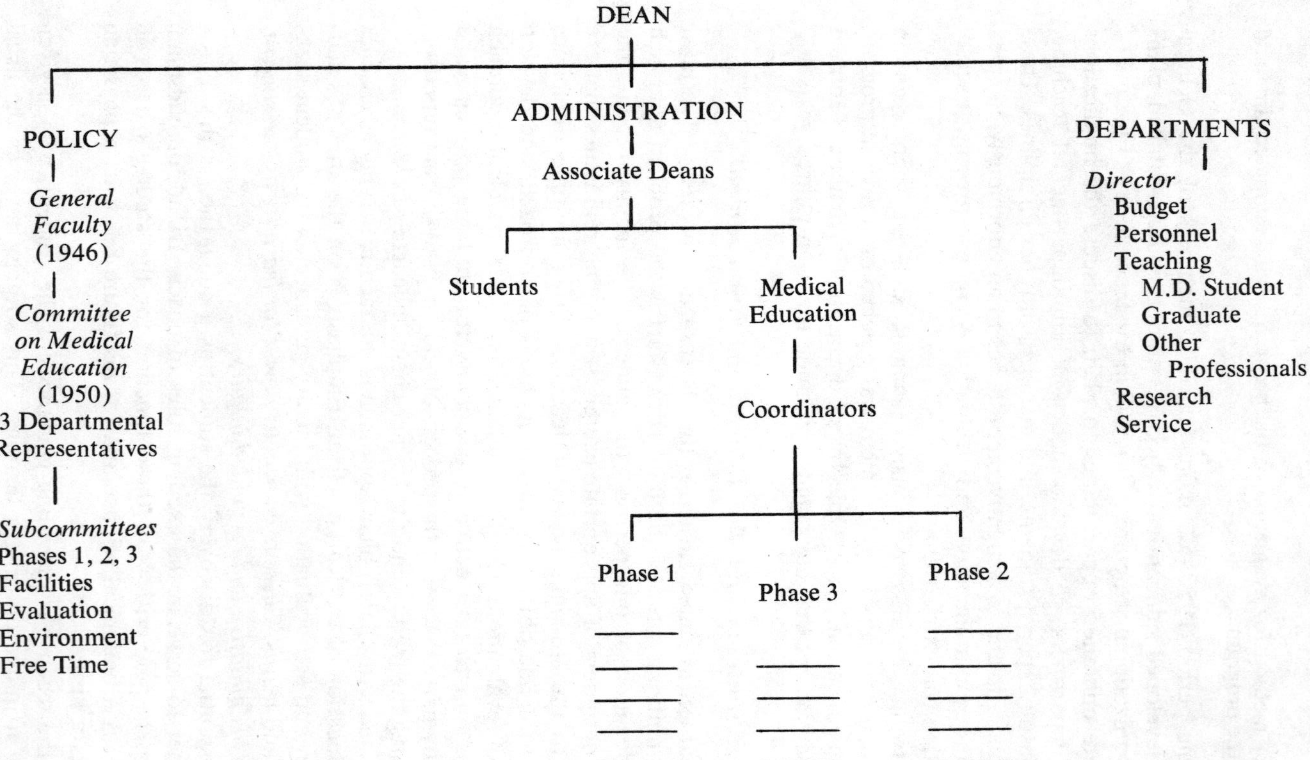

Figure 4. The organization of the faculty, administration, and departments for medical education at Western Reserve University.

role of the teacher as master; the responsibility is transferred, with clarity, to the student.

This strategy does not eliminate all forms of pedagogy, however, since it may be best for the client to use the indicated forms described below. The difference may well be the way pedagogy is used, that is, consciously and without loss of the learner's responsibility.

In the initial planning of "Objectives Concerning Curriculum" the treatment of the student was outlined as a key precept of the whole program. Understandably, these statements are a series of hypotheses that need criticism and experimental trial, studies of feasibility and cost, and definitive evaluation. But these strategies are known to excite the student as he finds satisfaction in accomplishing new tasks.

How to Get There?

How is the student to arrive at his goals in medicine? While the answers certainly are complex, they are essential for student learning and are designed to help him find his own way in the processes of learning. Four postulates are placed first in all these considerations since they appear to be the primary requisites to maintaining and enhancing the educational environment in which the learner-client can take the initiative in his own learning.

Postulate No. 1. Caring about the student leads inevitably to considering the individual career of each person, advising him, and showing concern for his readiness, plans, and progress. It is the human and personal touch that rediscovers the client daily and recognizes each one's uniqueness. Caring about the student makes his effort meaningful to him and to his career, and sets up a relationship of faculty and student as colleagues. Thousands of curricular hours may be vitiated if there is no appropriate way of relating the many educational opportunities to the student as a person. The objectives seem clear enough: to define the needs of the student in a continuing and meaningful manner, as done by a coach, a colleague, an ombudsman, and a counselor. This can be non-threatening to the student and help him *learn how to learn* on his own.

How to accomplish this nonthreatening coaching role with the learner-client has not yet been sufficiently worked out by any of us. We must try, first, however, to define this strategy of working with the student and then develop appropriate tactics. There is only one ready solution, and that is for all faculty members to participate as coaches and ombudsmen in their individual contacts with the student. We all need to learn how to do this, and we can in time.

Some questions need to be asked. Is the process of caring about the student as important as indicated here? Is it first in the hierarchy of the four postulates? Can we define more clearly our premises of working with the student, and our tactics? Can the monitoring of learning be done by faculty, by house officers, or by students? How can students as clients contribute to this process by stating their needs? What research in indicated?

Postulate No. 2. What is important is a series of statements by the faculty of what they believe is essential in terms of factual knowledge, of the use of that knowledge, of performance in learning, of problem solving, of skills, and of patient care. Nothing is more challenging to the faculty than attempting to answer the question: "What is important for the student to learn?" This inevitably involves the everchanging content of medicine, skills, use of information, and patient care. It also involves the skill of the learner in approaching problems, in learning itself, in being aware of what he does know, and, especially, in how to find that out. In some ways we are fortunate because the demands of medical science, clinical science, and patient care are so great that we must choose the imperatives for the learner-client in each of the above needs. There is no escaping selection, statements of priorities, and clear-cut objectives that can be understood by the student.

As a result of the overwhelming need for imperatives, the concept of a core curriculum has been developed that offers a realistic approach and is a tactical translation of strategic planning. Since *time* is the devastatingly limiting factor, faculty and students are challenged continually to keep revising the core approach to fit the realistic pedagogical approaches to the self-education of the learner-client. This is difficult, indeed, but it has been done and can be done by providing examples in biology and in medicine. A strategy of careful selection must precede the tactic of establishing a core curriculum, otherwise many bizarre results may come from unbalanced programs that deteriorate into a mass of didactic exercises, loss of the problem approach, and loss of self-education. Pedagogy then reverts to a primitive kind that quickly stuffs and stultifies the learner, who is given answers and not questions.

The challenge is real and difficult but the rewards to learning and to medicine can be enormous. Succinctly: Can we plan a series of imperatives, give them to the student as his guide, and still preserve his initiative in learning on his own with satisfaction?

Postulate No. 3. Guidance of the student. It is probably wise to regard all pedagogical methods as guidance systems designed to help the client accomplish his *mission,* whatever it may be. Thus the guidance is *for* the learner, to be used by him for a task; guidance is, by definition, not done *to* the student. The learner should therefore have a series of systems he employs according to the imperatives set by the faculty. Once again we must begin with treatment of the student and the educational environment, and then proceed stepwise.

Methods in the teaching-learning process represent a cafeteria of approaches available to fill the learner-client's needs. Since multiple methods are available, it behooves faculty and students to choose with wisdom, economy, and availability, but the student should always be able to *predict* what he can learn through a guidance system, the amount of time required, the kind of approach, the coverage, method of presentation, method of participation, method of self-evaluation, and the kind of coaching to be expected from faculty or fellow students. He needs to know the take-home pay from his work.

Postulate No. 4. How Am I Doing? The system of evaluation greatly affects the educational environment by the emphasis placed on examinations and on the monitoring of clinical performance. Evaluation undoubtedly holds the key to many aspects of education in the future. By evaluation the faculty shows what is expected, how the student is to be coached in formative evaluation, and how he is to be qualified in summative evaluation of his skills and knowledge of principles.

The student inevitably looks to the evaluation system to answer the question: "How am I doing?"; and the faculty looks to it to answer the question: "How have we done?" Our assessment systems are rudimentary indeed and bear little resemblance to our explicit goals of learning and performing the problem approach to learning itself, to medical science, and to patients' problems. Instead of trying to evaluate the student's ability to approach a problem, to select relevant information, to correlate and integrate information, and to offer a proposed solution, we often treat him as a memory bank to be tested for retention of enormous detail. We are attempting to develop appropriate systems to monitor the performance of the student as he studies the patient and as he applies the problem approach to the care of the patient.

But it is not enough just to criticize. From a positive point of view we recognize that the evaluation processes set much of the learning environment for the student—these processes literally guide the student in his approach to learning. Systems are needed that will allow the student to qualify, when ready, in a whole host of courses, skills, and performances based on the needs of medicine and medical care.

It behooves us, then, to devise evaluation systems that will reveal the student's effectiveness in learning on his own, and his ability to apply the problem approach to medicine and to patients. We have hardly begun such a scientific approach to the coaching of each student in the profession of medicine. We have little bench research on qualifying the student based on his true needs in medicine. We are at a beginning in evaluation.

From these four postulates the student will realize the key words that can create order out of disorder, namely, *responsibility* of the student to participate on his own, and authority to do so. This has already occurred successfully by treating the student as a colleague who is held responsible for learning on his own.

Selection of Subjects and Skills

In 1952 the School of Medicine at Western Reserve University selected subjects and skills that have remained relatively unchanged for twenty-one years. The interdisciplinary teaching by subject committees continues (Table 2). Phases 1 and 2 extend for one year each and are taught six mornings a week; in the same periods the students have three half-days free and two half-days for options. The five core clerkships and ambulatory clerkships take two months each, or twelve months. There are eight months for electives in the fourth year. Also shown in Table 2 are the choices of options in Phases 1, 2, and 3.

Guidance of the Learner

From long analysis of teaching, of learning, of pedagogy, and of the many aspects of education, there seems to be one oversimplification that may be stated before taking up the complex: the only kind of education for the learner is self-education; all educational resources and all our evaluations are guidance systems to help the learner.

The needs of the individual student to learn on his own are therefore paramount. The curriculum is about the eighth step in the hierarchy of approaches to self-education; qualifying when ready is about the tenth step in medical education.

We must now face the multiple systems for guidance of the student as he educates himself. We have recognized many systems, and each is important, each has a sequential position, and none can be discarded. An attempt is made in Table 3 to list these topics from the beginning, that is, learning skills. We are all aware of the essential nature of reading ability, communication, understanding, the problem approach, and use

TABLE 2. QUALIFYING AS A PHYSICIAN (CORE) AND OPTIONS

QUALIFYING AS A PHYSICIAN	CAREER PLANNING, OPTIONS, ELECTIVES, FREE TIME

Phase 1 (Six half-days)
  Cell biology
  Differentiated-cell
  Metabolism
  Cardiovascular-pulmonary-renal
  Tissue injury and disease
  Mechanisms of infection
  Biostatistics
  Clinical science
	(Three half-days free;
	two half-days options *)
	
	Phases 1 and 2 only
	  Non-medical school departments
	    11 courses—28 enrolled
	  Graduate (Medical School)
	    53 courses—265 enrolled

Phase 2 (Six half-days)
  Musculoskeletal
  Nervous system
  Reproductive biology
  Mind
  Gastrointestinal
  Hematology
  Biostatistics
  Integument
  Legal medicine
  Respiratory
  Cardiovascular
  Urinary tract
  Endocrine
  Clinical science
	Seminar Courses
	  Health care
	  Environmental medicine
	    7 courses—90 enrolled
	
	Practitioners and students
	  7 courses—119 enrolled
	Pharmacology
	  1 course—94 enrolled
	Resuscitation and emergency care
	  1 course—95 enrolled
	Human sexuality
	  1 course—25 enrolled
	Physiology: cardiovascular-
	    pulmonary-renal

Phase 3 (Full time)
  Core clerkships
  Medicine
  Pediatrics
  Psychiatry
  Reproductive biology
  Surgery
  Ambulatory
Curricular options (Full time)
  Clerkships
  Specialty subjects
  Biological subjects
  Research
  Extramural studies
  Vacation
	2 courses—20 enrolled
	Correlation gastroenterology
	  1 course—99 enrolled
	Interpersonal relations
	  1 course—48 enrolled
	Pulmonary pathophysiology
	  1 course—16 enrolled
	Computers and therapeutics
	  1 course—4 enrolled
	Mental development
	  1 course—9 enrolled

* Includes years 1968–1972 for Phases 1 and 2 only.

TABLE 3. GUIDANCE SYSTEMS FOR SELF-EDUCATION

LEARNING SKILLS	LEARNING MEDICINE
Reading Ability	Qualifying as a Physician
understanding	biological sciences
self-study	clinical sciences
self-learning	patient care
Communication	Career Options
written	nonmedical
verbal	biological, clinical
Understanding	health fields
questions	
problems	
Problem Approach	Systems of Communication
application principles	self-education units
correlation	written
association	audiovisual-written
analysis	computer
integration	lecture
Use of Literature	laboratory
study methods	conference, seminar
building own library	preceptorship
	problems
KNOWLEDGE	Evaluation Systems
	self-evaluation, formative
biology, chemistry,	auditing performance
physics, mathematics,	qualifying when ready
social sciences	summative

of the literature. Without these skills the student may be lost, waste enormous effort, and vitiate sophisticated programs. In the learning of medicine, qualifying in core studies, career options, systems of communication, and evaluation systems become evident as sequential and closely interrelated learning activities. There is no one system of guidance except self-education.

Curriculum

At last we have arrived at the curriculum, which is the development of a detailed tactical translation of the preceding strategic steps. The first definitive report of the program made in 1962 gave details of the curricular sequence and content.[7] Certain features are obvious. The subject committees continue, as do the clerkships. Options are 22 per cent; lectures 35 per cent; laboratories only 16 per cent (previously 31 per cent). The other activities such as conferences and clinical work in

Phases 1 and 2 are 22 per cent (down from 39 per cent). There is unmistakable delegation to the student of 45 per cent of the first two years as free time and as options. There are many whole class exercises, syllabi, lectures, and demonstrations. Students are not required at lectures or laboratory and take a large responsibility for learning on their own.

Description of the Program

Multiple publications have been issued about the program, as listed in the *Bulletin of the New York Academy of Medicine*.[8] It is a laborious process to present complex programs in a meaningful way. Also numerous staff meetings have been held for periods of one or more days to present programs, results, evaluations, and proposals. Communication is a difficult and time-consuming process because it requires dialogue to be effective.

Tactical Examples of the Educational Environment

For twenty-one years the students have taken *interim* examinations after subject committees, and *comprehensive* or *National Board* examinations at the end of each year. All results are anonymous except for those students judged to need tutorial help. There is neither grading nor ranking. There is a feedback system whereby students are selected to meet with the coordinator and faculty at regularly scheduled intervals to discuss their programs. Over the past summer three groups of students took the initiative to *develop presyllabus materials:* one on biological science, one on obstetrics in the family care program, and one on pediatrics. These are now used by the incoming class. *Tutorial programs* for first-year students in the biological sciences have been conducted by selected second-year students on a one-to-one basis for fifty-five members of the class of 1976, and, currently, for forty-five members of the class of 1977. These are formal, confidential arrangements made through the Phase I coordinator by matching the first-year student and the second-year tutor, and paying the tutor on an hourly basis for what may take him as much as four hours a week to review materials, understanding, and examinations. This is a practical and tactical solution for the learning needs of incoming students. The Phase I coordinator sends a newsletter to all his students at weekly or longer intervals to give them the "gossip of the Rialto" in medical education.

In Phases 1 and 2, each of which lasts one year, the student has core curriculum only six mornings a week, free time on three afternoons, and curricular options on two afternoons. Thus the student must choose his own program for 45 per cent of the time.

Evaluation of the Student and the Impact on Learning

Evaluation is a broad topic that probably holds the key to much of our understanding of education in the future as we attempt to proceed with the individual student in a human and scientific manner. Formative evaluation, evaluation of skills, summative evaluation, and examinations for licensure and for specialties are processes that relate so much to the educational environment of the student that their impact on learning must be considered at all times. The sequence of these multiple processes in evaluation of the learner makes a composite picture; the final objective is learning by the individual student or physician.

Fortunately we are in a period when the true *needs* can be considered of student learning, patient care, medical science, faculty participation, and community health. Each of these needs has been defined by William A. Anlyan in his twelve recommendations for the year 1985.[9] We are approaching the use of research methods for our many imperatives but we have not arrived.

Evaluation systems are rudimentary and bear little resemblance to the explicit needs of the student for information, the use of this information, and the approach to the patient. There is little to help the student as he seeks to assess his own learning. We seldom evaluate the student's performance in approaching a problem, finding information, and analyzing the results. We have not set up criteria and methods that allow the student to demonstrate mastery of a program and to qualify when ready. We do not yet have adequate nonpunitive systems to monitor the performance of the student as he applies skills and the problem-knowledge approach to the care of the patient.

From a positive point of view we recognize that the evaluation processes set the learning environment for the student, and drive him in much of his approach to learning. Systems are needed that will allow the student to qualify when ready in many courses, skills, and performances based on the needs of medicine and medical care. It behooves us, then, to devise evaluation systems that reveal the student's effectiveness in learning on his own and applying the problem approach to medicine and to patients. We have hardly begun an effective coaching system, a formative evaluation of performance, to help the student in medicine and medical care. Most summative evaluation is based not on criteria of learning for mastery of a subject or on a skill, but on a norm curve. We have not advanced the process of evaluation based on performance criteria that we need to develop. Summative examinations for qualifying in medical school and for licensure are being revised to fit the true needs of the learner in medicine and medical care. This will be difficult; we are only at a beginning, and we require the same continuing creative research inquiry in this field of evaluation of performance as we require in others. We need bench research of a cooperative nature with schools of the health sciences, educators, evaluation experts, the National Board of Medical Examiners (NBME), and licensing boards for medicine and specialties.

It is encouraging that increasing inquiry into education is being advanced to accomplish the objectives of the current report of the Committee on Goals and Priorities of the NBME:

> The central theme of this report is the evaluation of the medical professional throughout the educational continuum, from entrance into medical school to retirement from practice. Because evaluation is an integral component of education, certification and licensure, their interrelationships are also basic considerations of this report.[10]

In the same report the committee finds that:

> The NBME, aware of its unintended influence on the educational process, has over the past decade initiated changes in design, content, and scoring of certifying examinations to provide increased relevance and flexibility in response to changes in medical education.[11]

The challenge before the NBME is recognized by them as enormous. Mahoney and Englehardt have produced a lucid, critical summary of our needs in evaluation for health education, bringing together the needs of the learner and proposed ways of evaluating the effectiveness of education.[12] Examining bodies, including faculties, usually neglect some of the key skills in their systems of evaluation because they are hard to test. How well, for example, do we examine each student's skills in self-education, in the problem approach, in attaining and utilizing knowledge on his own?

At Case Western Reserve University the process of evaluation is in continuing debate and probably will remain so forever because of the need for valid information to guide the student and the faculty. The policy of nongrading and nonranking students is, however, in its twenty-second year. There is today enthusiastic spontaneous tutoring by classmates on an extensive basis. There is no competition for class standing, but a keen interest in learning.

It is our opinion that *as a student is evaluated so will he study*. We must raise our sights, then, to recognize that we are educating a health professional to learn for the rest of his life how to approach problems, to know his limitations, and how to resolve them. As a physician there certainly is a need to qualify at different stages of acquisition of knowledge and skills. It is encouraging that inquiry has now turned to helping the learner know where he is and how he is doing—these are the roles of the teacher and of the certifying groups who must all keep pace with change.

Significance and Limitations

Establishing the significance and limitations of a program of medical education or of a curriculum is difficult indeed because of the nature of the process of learning required for a lifelong profession. We do not have adequate criteria for measurement as yet, and must rely on description, on responses of the learner, and on performance of graduates in their careers. We are dedicated to following the careers of graduates to learn how they evaluate their education. Betty H. Mawardi has studied graduates of the classes of 1935–1945, and is embarking on a study of those under the revised program, the classes of 1956–1965. Our concept of the individual career of each student serving as the unit for education and for evaluation is shown in Figure 5.

We need interschool cooperation for such evaluation so we may learn from the unique programs and their effects on graduates as observed at McMaster, Duke, Pennsylvania, and Washington, to mention only a few. Concerning the curriculum, it is obvious that many variations are resulting from the investigative approaches outlined here. Interdepartmental planning by general faculty organizations has been recommended in a report of the Association of American Medical Colleges.[13] Interdepartmental teaching by subjects is conducted in twenty-eight schools for the basic sciences and in forty-three schools for the systems of the body. There are regular three-year programs in nineteen schools, optional three-year programs in thirty-one, and elective fourth-year plans in sixty-five. Thirty-two schools rank their students. None of us can interpret the results of our efforts as yet, but students and faculty are developing better systems for reporting to one another during the education process, and there is real hope for a continuing study of graduates. Guidance systems are improving as we appear to delegate more successfully to the student the responsibility and authority for self-education of worthwhile problems in the biological sciences, clinical sciences, and patient care. But the specter of more lectures and more regimentation

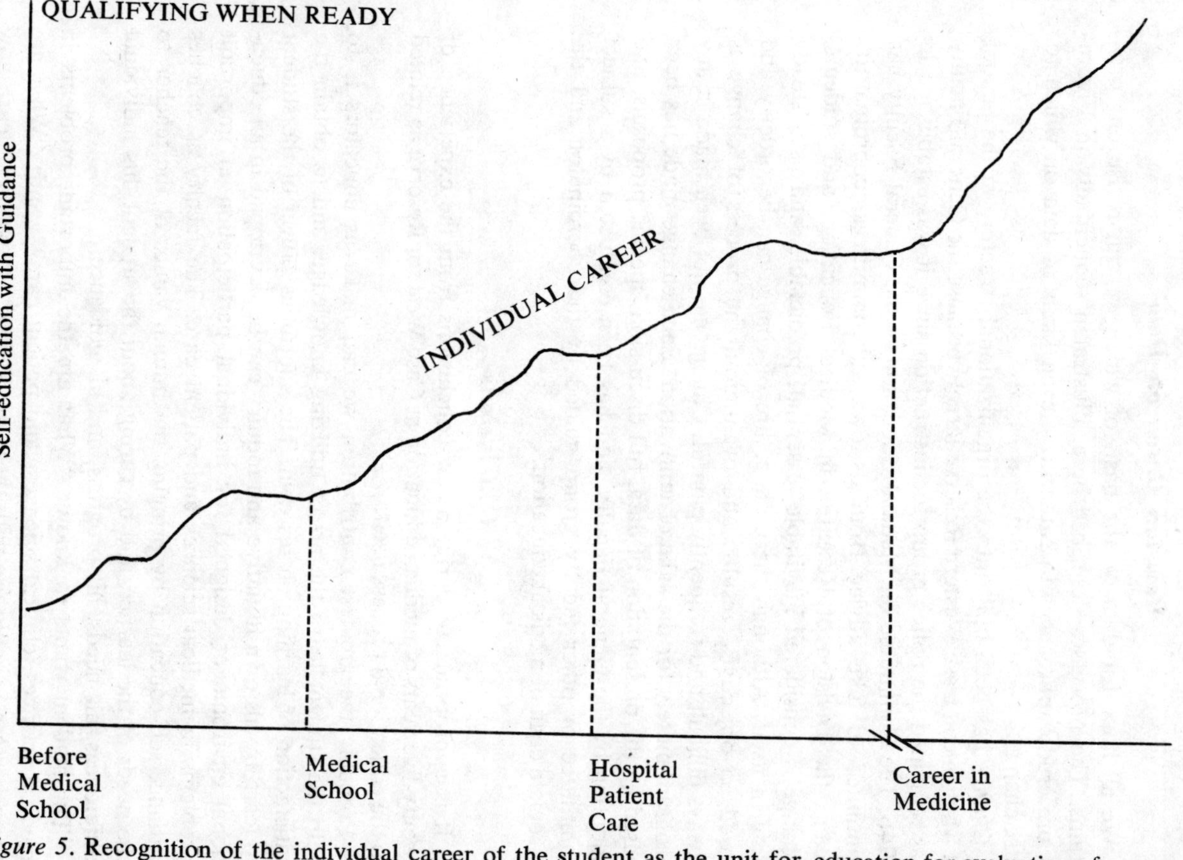

Figure 5. Recognition of the individual career of the student as the unit for education for evaluation of career.

is always a problem with the increase in knowledge, the expanding student enrollments, and decreasing funds.

Revision Based on Data

We all hope for data on the basis of which we can revise our programs. The processes described here, illustrated sequentially in Figure 6, are crude approaches to gathering eleven kinds of data on which to base change.

Change has been built into the organizational structure in the School of Medicine at Case Western Reserve largely because the General Faculty has remained the policy group for instruction since it was established in 1946. The Committee on Medical Education of the General Faculty has continued to take advice from its own subcommittees, coordinators, deans, the Division of Research in Medical Education, and students. Change is considered inevitable, essential, predictable, and as a steady state in itself. Although the architecture of change can be orderly and based on objectives, results, and evaluation, the process of change is always difficult and frequently painful. Change has not been made in any program merely for the sake of innovation. Each change made has been based on the presentation of data, full discussion, specific proposals, and adoption by the General Faculty, and has been conducted by a subject committee or other faculty group guided by the coordinator and the associate dean of medical education.

Conclusions

It is appropriate to derive a few conclusions from the experience of twenty-five years of medical education at Case Western Reserve as related to the educational process itself.

Probably the greatest contribution we can make as educators is to help the student learn the art of utilizing knowledge, and to obtain the satisfaction of learning on his own. The enormous power of the student to learn by his own initiative and inquiry can be developed in an educational environment designed for individual participation in important problems in medical sciences and patient care. The faculty as coaches enhance self-education by bringing the human values of the teacher to the needs of the learner, and by caring about the student, his individual differences, his needs, his progress, and his guidance.

The student wants to know: "What are the important problems in medical sciences, clinical sciences, and patient care?"; and: "What are the processes I should use with these problems?" He wants to learn how

1. Changing needs of medicine
2. Role of biological sciences, clinical sciences, patient care
3. Development of overall plan
4. Faculty participation
5. Students—educational environment
6. Selection of subjects and skills
7. Guidance of the learner

8. Curriculum
9. Describing program

10. Evaluation
11. Significance, limitations
12. Revision based on data

Figure 6. Investigative processes in sequential approaches to the learner in medical education.

to apply his information in a meaningful way to important phenomena in medicine.

The student will seek information on how he is doing by means of self-evaluation. The faculty can work with the student to develop methods for self-evaluation. The faculty as coaches can guide the student as he takes responsibility to learn on his own. This guidance applies to the biological and clinical sciences, as well as to patient care.

The interim examinations are considered as formative evaluation, and the student who requires further coaching is identified. Summative evaluation is a qualifying exercise in which the student is asked to demonstrate, in an appropriate series of examinations, his basic knowledge, his use of knowledge in problems, and his use of skills and knowledge with patients. Identification is made if the student does not qualify.

Many of the faculty who act as "guides" to the students come to understand the individual student—how he differs from others, his aptitudes, his interests, and his performance in the biological and clinical sciences and in patient care. This knowledge of the student is identified by many faculty members who serve as counselors, and is helpful when recommending the student for further training. The student's performance should be reviewed with him. While there is no ranking in class, there are serial evaluations of the performance of each student, and differences that are meaningful to his career and to his future contribution to medicine are reported to him.

NOTES

1. W. H. Glazier, "The Task of Medicine," *Scientific American* 228, no. 4 (April 1973).

2. A. Flexner, *Medical Education in the United States and Canada,* Report of the Carnegie Foundation for the Advancement of Teaching (Boston: The Merrymount Press, 1910).

3. Ralph W. Tyler, "Basic Principles of Curriculum and Instruction," *Syllabus for Education* 360 (University of Chicago, January 1950): p. 83.

4. T. H. Ham, "The Approaches of the Faculty to Medical Education at Western Reserve University," *Journal of Medical Education* 34 (1959): 1163–74.

5. Ibid.

6. "Objectives Concerning Curriculum," Draft No. 11, Approved by the General Faculty, Western Reserve University, April 10, 1951.

7. T. H. Ham, "Medical Progress, Medical Education at Western Reserve University: A Progress Report for Sixteen Years, 1946–62," *New England Journal of Medicine* 267 (1962): 868–74, 916–32.

8. ———, "Research in Medical Education, Participation of Faculty and Students," *Bulletin of the New York Academy of Medicine* 128 (1965): 501–508.

9. W. A. Anlyan "Chairman's Address, 1985," *Journal of Medical Education* 46 no. 11 (1971): 917.

10. *Evaluation in the Continuum of Medical Education* (Philadelphia: National Board of Medical Examiners, June 1973): 83.

11. Ibid., p. 31.

12. M. A. Mahoney and S. L. Englehardt, "Toward an Evaluation System for Health Education," *Pharos* (April 1973): 62–66.

13. W. N. Hubbrd, Jr., J. A. Gronvall, and G. A. DeMuth, eds., "The Medical School Curriculum," *Journal of Medical Education* 45, no. 11, pt. 2 (1970): 1–61.

General Discussion

The principal problem in medical schools in the United States today may be the inability of faculty to communicate with each other, rather than their inability to communicate with the students. Concepts taught in basic science departments may never be referred to by clinical departments.

It was emphasized that good clinician-teachers must have a reasonable knowledge of the basic sciences in order to explain disease mechanisms. There was general agreement that pediatrics usually leads the clinical departments in relating to the basic sciences.

Clinical pathology conferences do not contribute significantly to the integration of the basic and clinical sciences because they are "on-stage performances"—they are actually inferior teaching exercises.

One failure in the "grand design" at Western Reserve was the attempt to bring students back to the basic sciences in the final year. Other failures included student research projects, a continuity clinic, and family care programs.

Once the medical student becomes involved with clinical care, he or she usually shows no interest in programs designed to integrate knowledge in the basic sciences with clinical experience.

The National Board of Medical Examiners: Changing Policies and Practices

J. D. MYERS

The National Board of Medical Examiners (NBME) came into being shortly after the release of the Flexner Report. Its objective was to provide standardized examinations of high quality for purposes of medical licensure. For years the examinations have been administered in three parts: Part I at the end of training in the basic medical sciences; Part II upon formal completion of the medical school curriculum; and Part III after finishing the internship or the first year of graduate medical education. In spite of considerable variations in the pattern of medical education, there has been very little change in the format of the overall examinations over the years. On the other hand, the examinations have been changed by the introduction of new testing techniques such as multiple choice questions and patient management problems (PMPs).

The Committee on Goals and Priorities

It has always been the philosophy of the NBME that it should be responsive to the pattern of and changes in medical education. While the NBME should not determine the nature of medical education, it is widely recognized that any examination procedure has an inevitable influence on the educational program it tests. In recent years the NBME has become increasingly concerned that its set format of a tripartite examination is no longer in keeping with modern medical education. In addition its activities have extended into many other fields than merely providing examinations for medical licensure. Several years ago a Committee on Goals and Priorities was appointed to advise the NBME regarding its future activities; the report of that committee is now generally available.

The report of the committee (the GAP report) proposed some very definite modifications in the testing of medical students on their accomplishments in the basic medical sciences. It has been recognized that, increasingly, students are coming to medical school with distinctly varied backgrounds. In recent years many medical schools have directed that clinical studies be introduced earlier in the curriculum. Using the biological systems approach, the Case Western Reserve—or vertical—system of medical education teaches the preclinical and clinical sciences in combination in an organized and integrated sequence. Following a similar pattern, in many medical schools the interface between the basic medical sciences and the clinical medical sciences has become blurred. Some curricula have been shortened, while other institutions have adopted a combined college and medical school curriculum, usually of six years' duration.

Elective time has been built into curricula so that in many schools it now occupies all or a major portion of the fourth year of medical education. Some students, but disappointingly few, take electives in the basic medical sciences during the latter part of their education.

With the rapid expansion of medical knowledge during the last twenty-five years, the basic medical sciences are becoming increasingly pertinent to the understanding of clinical medicine; the most successful clinical investigators at the present time are also schooled in one or the other of the basic sciences.

The Qualifying A Examination

All of the above factors have led to a situation in which one can no longer say that the first half of medical education is devoted to the basic medical sciences and the latter half to clinical subjects. This means that the traditional Parts I and II of the NBME examinations are no longer appropriate and in keeping with modern medical education. In fact for several years now the NBME has allowed students to sit for Part I or Part II in whatever sequence they prefer; thus Part II may be taken before Part I. The GAP report recommended that Parts I and II be combined into a new examination, termed the "Qualifying A Examination," to be administered at the end of formal medical education. Discussion by members of the NBME has emphasized that the Qualifying A Examination should contain a strong component of basic science material. In many instances basic medical science information can be utilized in questions that have clinical import, but it is not intended that the basic science component of the examination be totally clinically oriented. It is believed that examining for the basic medical sciences at the end of the medical school experience has the advantage of helping to maintain the student's interest in them after his three- or four-year curriculum. Today the successful completion of Part I often tends to depress interest in the basic sciences, and this of course is not compatible with good medical education.

It is not expected that there will be identifiable sections in the Qualifying A Examination for each of the basic medical sciences. There will, however, be adequate material in each, and according to present plans it will be possible to extract a subscore from the overall examination for each basic medical science. This will serve to provide feedback to medical school faculties regarding the performance of their students in each area.

The Qualifying A Examination is to be accompanied by an intramural evaluation of the students by their own institutions. In this regard the NBME has expectations that it may be able to provide new forms of achievement examinations that the schools can use for part of their internal evaluations. The NBME will continue to make examinations available in each of the individual preclinical and clinical medical areas, and schools can use these as they see fit to measure achievement. In fact, a school may, if it wishes, combine the several individual examinations in the basic medical sciences to form an equivalent of the current Part I. Test committees will be retained in their present form to draw up these individual achievement examinations as well as to devise the Qualifying A Examination. The committees will thus continue largely as now constituted in both the basic medical sciences and clinical disciplines.

It is planned to form a Council on Undergraduate Medical Evaluation to supervise these activities in the field of undergraduate medical education for the NBME. The council will have strong representation from the basic medical sciences as well as from clinical areas, and, further, representation from the Federation of State Boards, so that they may use the Qualifying A Examination in granting a license to the new medical graduate to participate in patient care under supervision during his internship and residency years.

Some medical schools have already criticized the abandonment of Part I as an individual examination. They have expressed the opinion that this downgrades the basic medical sciences, and that the schools continue to need an external examination confined to the basic medical sciences, which includes the clout of licensure. As mentioned earlier, the NBME will continue to provide achievement examinations in individual basic medical sciences or combinations thereof. It hopes that by proper use of achievement examinations during the medical school curriculum, and by proper methods of intramural evaluation, the loss of "incentive" on the part of the medical student, vis-a-vis licensure, will not be significant.

A secondary reason for combining Parts I and II into the Qualifying A Examination is to decrease the number of formal examinations to which undergraduate and graduate medical students are exposed. The multiple administration of examinations is particularly noticeable during the graduate years, with the National Board's Part III, various in-training examinations during house staff periods, and the like.

A strength of the Qualifying A procedure is that foreign medical graduates wishing to enter medical programs in the United States would be required to pass the examination, and thereby, at least from the external examination standpoint, their qualifications would be comparable to those of American graduates.

The NBME will continue to provide advanced placement examinations for candidates wishing to enter the medical school educational process at a later period.

In summary, the National Board of Medical Examiners is currently of the mind that the new arrangements I have described will lead to a strengthening of the position of the basic medical sciences in the medical school examination process. Certainly there is no desire to downgrade the importance of the basic medical sciences in medical education or in examinations, and all efforts will be exerted to see that this is not done.

General Discussion

Although the newly graduated physician is not ready for the independent practice of medicine, several states have no requirements for further training. Probably no more than 1 or 2 per cent of graduates move directly into practice from medical school, but even 100 is too many. All new physicians should undergo a period of postgraduate training.

The security of tests for foreign medical graduates is a major problem. As an example, Manila is on the other side of the International Date Line, and the trans-Pacific telephone is used by graduates who have just completed the ECFMG examination in Manila to relay the questions to their friends in the United States the day before they take the examination.

The Undergraduate Program in
Human Biology at Stanford University*

NORMAN KRETCHMER

The undergraduate Program in Human Biology was officially launched at Stanford University in 1969, during a time of great student unrest. The program was designed to utilize faculty from various university schools and disparate departments, and to present a cooperative curriculum leading to an A.B. degree in human biology. The stated objective was to attempt to integrate into the educational effort all aspects of man and his environment. This presentation will attempt to examine the content and attainment of the goals of the program by blending the attitudes of students and faculty with my personal prejudices.

By the usual academic measurements—number of students registered; rate of placement of students in professional and graduate schools; faculty plaudits; financial stability; and growth—the program has been a resounding success. Since it has been called a seminal experiment within a university, however, it is fitting to present the substantive content of the program for judgment by peers. Within this designation I unequivocally include the student, who is the ultimate utilizer of the curriculum.

The curriculum in human biology consists of a regular university major involving about sixty-five quarter credits and leading to a bachelor's degree (Appendix B). The core program is obligatory and constitutes half of the credits; the other half consists of courses taken in both biological and social sciences, either from the advanced offerings within the program or from courses at a similar academic level in any department on campus. The advanced offerings can generally be viewed as falling into a category designated either as a natural or a social science.

Student response to the program was immediate and dramatic, and it now has the third largest student body in the School of Humani-

* We wish to acknowledge with deep appreciation the substantial support the Ford Foundation has provided for the Program in Human Biology.

ties and Sciences. A major area of study must be declared by the student
at the beginning of the junior year: in terms of student selection, psychology is first; biology a close second; and English fourth (Table 1).
More than half of the students in the School of Humanities and Sciences
have indicated a major field of interest, and 35 to 37 per cent are in
psychology, biology, human biology, or English. In 1972, 16 per cent of
all students could be identified as desiring a career in medicine—
27 per cent of the known majors.

The student population in the core courses in human biology gives
better insight into the program. The first course, emergence of man, is
often taken by students, usually freshmen, who are fulfilling their science
area requirement and who are not interested in continuing in the major
(Table 2). A better indicator of serious interest in the program are the
figures for the second and third courses, for students are required to take
the natural and social science sequences simultaneously. The core program is required of all majors and must be taken in sequence. The natural science portion introduces students to evolution, to cell biology and
neurobiology, to human development and homeostasis, and finally to
population biology. These subjects are intended to converge with the
social science courses on primate behavior, developmental psychology,
social and economic systems, and psychobiology.

There are indications of a decline in student enrollment in the program, a possibility that can be judged with more validity after tabulation
of the number of majors in human biology in the fall of 1974. As the
core progresses the number of students declines as they become increasingly obligated to fulfill the complete major.

TABLE 1. STUDENTS IN DECLARED MAJORS AT STANFORD UNIVERSITY

Major Department	1970	1971	1972	1973
Biology	282 (3)*	352 (3)	405 (1)	432 (2)
Psychology	384 (2)	390 (1)	384 (2)	434 (1)
English	397 (1)	359 (2)	272 (4)	218 (4)
Human Biology	113 (4)	292 (4)	324 (3)	320 (3)
Major declared	3,594	3,762	3,934	3,811
Major undeclared	2,709	2,669	2,478	2,626
Total undergraduates	6,303	6,431	6,412	6,437
Premedical students			1,046	

* Ranks of preference among majors are in parentheses

TABLE 2. PROGRAM IN HUMAN BIOLOGY AT STANFORD UNIVERSITY: CORE COURSES

Core Courses	Students				
	1969	1970	1971	1972	1973
Emergence of Man	438	409	630	393	—
Cells, Organisms, Societies	—	306	267	287	193
Evolution of Human Behavior	—	307	266	286	185
Man as an Organism	—	231	235	239	—
Development of Behavior	—	229	222	232	—
Biology of Populations	—	202	210	232	—
Social Organization of Man	—	192	204	225	—

The original objective was to design advanced courses that demonstrate the interrelationships of social and natural sciences (Table 3). A few of the courses satisfy this demand: human nutrition, human genetics, birth control, and psychobiology are prime examples; other courses meet very specific needs: biochemistry, political processes, and manmade environment. The faculty teaching this array of courses derive not only from those associated with the Human Biology Program, but from other university departments. Courses are added by joint faculty-student decision.

The number of new students entering the program who are interested in studying the premedical health sciences has climbed from 22 per cent to 52 per cent since 1970. In part this growth of interest in medicine reflects an increase in the stated desire of entering students to seek careers in medicine and the health sciences: 16 to 25 per cent of the present student body has medicine as a career goal. It is shocking to think that 50 to 75 per cent of the students in a liberal university such as Stanford have indicated that they will go into medicine, law, or engineering. As a number of students have put it: "That's where the jobs are."

We have had the opportunity to survey the most recent graduates of the program (Table 4). Of the ninety-three students who replied to a questionnaire, sixty-three had applied to professional schools and fifty

TABLE 3. PROGRAM IN HUMAN BIOLOGY AT STANFORD UNIVERSITY:
ADVANCED COURSES

Advanced Courses: Natural Science

Introduction to Biochemistry (50)*
Human Genetics (60)
Human Nutrition (89)
Energy and Energy Resources (59)
Natural History of Infections (63)
Human Health as Human Ecology (10)

Advanced Courses: Social and Behavioral Sciences

Political Processes (22)
Manmade Environment (18)
Comparative Survival Strategies (5)
Transformations in Prehistory (19)
Biosocial Aspects of Birth Control (29)
Primate Behavior (72)
Topics in Psychobiology (28)
Human Aggressiveness (17)

---

* Number of students enrolled are in parentheses

TABLE 4. FOLLOW-UP OF GRADUATES OF THE STANFORD PROGRAM
IN HUMAN BIOLOGY, 1973

Number of students	127
Replies received	93 (74%)
49 (52%) entered program as preprofessional preparation (majority premedical)	
22 (23%) also completed another major	
63 (67%) applied to graduate or professional schools (50 [79%] were accepted)	
Plans for 1973–74	
Medical school	30
Graduate school	13
Other professional schools (law, etc.)	7
Jobs	16
Travel, research, etc.	16
Undecided	11

---

had been accepted. Forty-nine of the students indicated an interest in a professional career, and thirty had been accepted by medical schools. This is a minimum figure, since the exact number who applied to medical school is not known.

Student Reaction

Although interest may be waning, the program has been a very popular major with undergraduate students at Stanford. Its approach to a number of careers differs from the standard major. In the words of the students, in their own handbook:

> People have been asking the question, "What is human biology?" for as long as the program has been around. The variety of responses to this question may be an indication that there is no single answer. Human Biology is different for each individual who is connected with the program. The flexibility which allows the program to be shaped to the individual interests and needs of every undergraduate major is highly desirable.

Students in the program are dedicated, and they have ample opportunity to participate in its development through consultations, service on committees, formalized student critique, and social hours. A wealth of criticisms and suggestions has derived from these activities. Another comment from the student handbook gives an apt summary:

> If there have been contradictions between the original goals of the Human Biology Program and the present-day realities, it is in part a reflection of the evolution of the program. It evaluates itself each year in light of student criticism and approval. Weak points have been spotted and they are being strengthened. The strong points are recognized and reincorporated into the program as it develops anew each year. These initial years have seen much change and adjustment. The program has proven itself capable of standing on its own feet, but there are still problems to be met. A particularly big challenge for the program is the integration of fields which traditionally have been studied separately.

Since half of the students in the program are interested in a medical career, there has been constant criticism from the general student body that the curriculum is focused primarily on the needs of the premedical student. This admonishment relates to a concentration of the course material on the human, and the tendency for professors from the medical school to adhere to their specialty and to utilize clinical examples in their teaching.

On the other hand, when the teaching takes cognizance of those students interested in the law or the behavioral or social sciences, the accusation of "soft science" is hurled at the program. To the purist, this allegation probably has some validity. The serious student of psychology, biology, or anthropology may rightfully be dissatisfied with the program.

Another difficulty, and a major one, is a lack of convergence of the natural and social sciences. Multidisciplinary programs are fraught with pitfalls, and this one is no exception. They require not only a blending or convergence of course material, but a willingness and capacity for constant communication among the faculty in the various disciplines. Neither requirement can be attained with facility.

Because of the size of the classes, students express feelings of isolation, of being treated impersonally. This attitude applies not only to the large core courses, but to the bigger-than-desirable population in the advanced courses. Because of the virtual lack of restriction on enrollment, we and the students suffer from the popularity of the program.

Another problem that is somewhat reflected to the student is that we, the faculty, are also isolated one from another. Most of the program's faculty members have teaching responsibilities in their home departments as well as in the program, and their investigative and administrative loyalties lie with the former. In actuality, an itinerate group of very senior faculty teach the core program. While the fact that renowned senior faculty teach beginning undergraduates certainly does make the program more attractive, at the same time it creates distinctive problems. For secular, academic, and administrative reasons it is very difficult to recruit younger faculty members who are making great efforts to foster careers in their particular area of interest. The younger faculty who have become involved are really devoted to the students and to the program. A number of suggestions could be made for overcoming these difficulties, but since the specific concern of this presentation is the examination of premedical and medical education, I will say no more on the subject.

Relationship to the Medical School

The Human Biology Program could readily be intertwined with the curriculum of the School of Medicine. A number of undergraduate students in their senior year are actually attending several basic science courses in the medical school. Thus they are blending the two levels of medical education, and if they enter a medical school that recognizes advanced placement, they will have an advantage not available to students taking a standard curriculum. The program is therefore a "natural" for creating an integrated undergraduate-medical school curriculum, by providing a smooth mechanism by which to shorten and tighten the lengthy training for the M.D. degree.

To idealize the above proposal requires the continued participation of medical school and undergraduate faculty, and a recognition of the advanced education offered by many high schools in the nation.

Our core program could be given in one year, and all the premedical requirements could be completed readily in three years. There are overlapping fields, such as biochemistry and physiology, in which even now upper-division undergraduates are active participants. Other subjects taught within the environs of the medical school deserve as hard an appraisal as anatomy has received over the past few years. There are also a number of fields that could be taught jointly more effectively—anatomy-physiology-pharmacology, and obstetrics-pediatrics, for example. A considerable revision has and can be made in the medical school curriculum, even to the extent of a critical appraisal of the value of the extensive elective programs encountered in most medical schools. The objective is to shorten the traditional amount of time spent by the student on his medical education.

The faculty of a professional school need to be fully aware of the curricular changes on the undergraduate campus. The program described here, because it leads to the baccalaureate degree, is perhaps more ambitious than most, but it is a fair example of the general response of many colleges and universities to the concern of today's students with personal and societal issues. The interdisciplinary trend in the undergraduate curriculum will inevitably have an effect on the medical school curriculum. It will be necessary for medical school admissions committees to learn to evaluate an applicant with a somewhat different course pattern, but with the substantial asset of a greater awareness of man as an organism.

The recent revisions in medical school curricula throughout the nation have taken a variety of forms: expansion of elective opportunities, reduction of the length of certain courses; incorporation of clinical experiences earlier in the educational sequence; consideration of incorporating the equivalent of the internship into the premedical years; and increase in the free time of students to pursue their own clinical or investigative interests. Some schools have even had the imagination and courage to eliminate traditional courses or to consolidate some that are closely related.

If a critical examination is made of the product of the medical school, the doctor, and what he or she has to contend with in modern society, the importance of the Program in Human Biology to medical training becomes apparent. The physician must be aware not only of the diseases of man, but, more important, of the meaning and preservation of health.

He must know how man interacts with man and society, both in physiological and behavioral terms. Every physician should understand fully the extent to which environmental factors relate to the maintenance of health and the pathogenesis of disease. The student entering medical school who has been exposed to the kind of curriculum I have described will not only be concerned about his fellow human beings, but will have an appreciation of man, his society, and his environment.

ॐ

General Discussion

The Stanford Program in Human Biology originated when a representative of the Ford Foundation proposed to the faculty that such a program should be established.

No special texts have been prepared for use in the program. The enrollment does not consist primarily of premedical students; about 10 per cent, for example, are headed for careers in law.

One criticism made by students is that the human biology program is a "watered down" biology curriculum, whereas some may prefer standard courses that afford better opportunities for admission to medical school. The importance was stressed of developing new and stimulating courses in programs in human biology. The problem of articulating programs with other professional schools in the universities has not been resolved.

The principal purpose of the Stanford program is to improve the quality of instruction in human biology; it does not have as its primary objective the preparation of students for advanced placement in medical school.

There has not been sufficient elapsed time to follow the career choices of students who have completed the program at Stanford.

The Program in Human Biology at Stanford University*

COURSES AND DEGREES 1973–74

ह∾

Committee in Charge: Donald Kennedy (Biological Sciences), *Chairman;* Sanford M. Dornbusch (Sociology); Paul R. Ehrlich (Biological Sciences); David A. Hamburg (Reed-Hodgson Professorship of Human Biology and Psychiatry); Albert Hastorf (Psychology); Norman Kretchmer (Pediatrics); Joshua Lederberg (Genetics); Colin S. Pittendrigh (Bing Professor of Human Biology and Biology); James L. Gibbs, *ex officio* (Dean of Undergraduate Studies).

Faculty: John E. Adams (Psychiatry); Albert J. Ammerman (Genetics); Keith Brodie (Psychiatry); Luigi L. Cavalli-Sforza (Genetics); Garth Collier (Civil Engineering); Peter Corning (Political Science); Carl Djerassi (Chemistry); Shirley Feldman (Psychology); Jane van Lawick-Goodall (Psychiatry); John G. Gurley (Economics); Nicholas J. Hoogenraad (Pediatrics); Herant Katchadourian (Psychiatry); Sidney Liebes, Jr. (Genetics); William V. Robertson (Pediatrics); Alberta A. Siegel (Psychiatry).

Student Members: Donna Anderson, Cynthia Clinkingsbeard, Laurie S. Gill, Robert Kaplan, Daniel J. McFarland, Anne M. Murphy, Iris M. Payne, Randall A. Yim.

Program Coordinator: Sophia C. Alway.

Statement of Purpose

This Program is an undergraduate major designed to encourage the convergence of natural and social science in the study of man. The Program is an interschool, interdepartmental major, utilizing not only those faculty and courses particularly created for the major, but also pertinent areas of instruction available throughout the University. It also is concerned with man as an organism, his adaptation to other men and to

* Reprinted, with permission, from the *Stanford University Bulletin.*

nature, his ability to control and to live with the environment, and the mechanism by which these factors relate to his biological and behavioral evolution.

This Program is a response to the need for knowledge of the complex relationship of man with nature, exemplified by the dilemmas of social policy in health and education, population problems, pollution of the environment, and conservation and development of resources. The Program is designed for the general education of policy makers and citizens. It is also a route to advanced study in the established natural and social sciences and related professions.

Offerings and Facilities

The Program is funded by a grant from the Ford Foundation and leads to an A.B. in Human Biology. The curriculum is designed for those students who desire a knowledge of biology, particularly of man, linked with knowledge of the behavioral sciences. The Program predominantly involves faculty from the School of Humanities and Sciences and the Medical School. Representatives from other Schools will also participate in the Program.

The coure of the Program for majors in Human Biology is the Fundamental Program. It consists of eight one-quarter courses required of all majors. The objective of these courses is to present a broad but rigorous overview of the biology and behavior of man in society. The core is the necessary academic basis for the more specialized and advanced offerings of the Program.

There is no graduate program in Human Biology, but students will be prepared for advanced training in biology, the behavioral and social sciences, medicine, law, or education, depending on their choice of advanced courses following the Fundamental Program.

The Office of the Program in Human Biology is located in Building 80 of the Inner Quad.

Program of Study

Bachelor of Arts

The degree of Bachelor of Arts in Human Biology will require approximately 60 to 65 units in the major. The Fundamental Program will consist of 33 units and will satisfy the University Distribution Requirements in the social sciences and the natural sciences. It is expected that, in addition, at least six advanced courses will be taken in fields related to the biological, social, or physical aspects of Human Biology. Detailed guidance should be sought at the office of the Program in Human Biology so that the program for the individual student can be designed to fit his or her particular needs and career goals.

Courses

Note: Students who have elected a major in Human Biology will be expected to take courses 1 through 6 in the Fundamental or "Core" Program. These courses must be taken for a grade by majors with the exception of the workshop. It is advised that the sequence be initiated in Spring Quarter of the Freshman year. Courses 1 through 4 are open to nonmajors; however, the A and B Series must be taken concurrently and in sequence by all students.

Fundamental Program

1. Evolution of Life and Emergence of Man—The question "what is life?" leads to a discussion of the nature of organisms, of organization in general, its dependence on information, and the central position of genetic and evolutionary theory in all biological sciences. A beginning is made in developing an understanding of the role of natural selection in molding the character of organisms and societies as self-reproducing entities adapted to the conditions in which they exist.

A major section of this course is a substantial treatment of Mendelian and population genetics. The nature/nurture problem is introduced as one of the most important contributions which the biologists as such can make to an understanding of man and political issues that beset him.

Metabolism in general, with principal emphasis on the energetics of the organism and traffic with the environment in material constituents, is given only brief treatment. The cell is studied as the simplest unit of living organization. The structure of its organelles is considered in terms of the functions they serve, especially in terms of the energy relations.

This introductory course is primarily concerned with broad outlines of the origin and history of life, with special emphasis on the evolution of the vertebrates and the primates. The quarter will close with a discussion of the biological uniqueness of man and his origins from the Australopithecines.

5 units, Spr (Pittendrigh) MTWThF 10

2A. Cells, Organisms, and Societies—The structural and functional prerequisites for life at various levels of organization are treated in this quarter in greater depth, i.e., cellular structure, molecular architecture, and the energetics of living systems. The character of intercellular communication in multicellular organisms, leading to the central nervous organization underlying behavior and finally to social organization in animal societies, provides the major theme for the course. Prerequisite: 1 or Biological Sciences 1; must be taken concurrently with 2B.

4 units, Aut (Kennedy, Staff) MWF 9

2B. Evolution of Human Behavior—This course views man as an organism with a long evolutionary history that has significance for understanding the behavior of contemporary man. Over millions of years, behavior patterns have evolved in relation to meeting survival requirements: food, shelter, defense, reproduction, preparation of offspring to cope with environmental conditions. Such adaptive patterns will be examined in different eras of human evolution: in nonhuman primates; in hunting-and-gathering societies; in agricultural societies. Attention will be given to: subsistence patterns; interpersonal and intergroup relations; stressful experiences and their physiological concomitants; sources of conflict and modes of conflict resolution.

4 units, Aut (Hamburg, Goodall) MWF 10

3A. Man as an Organism—The intra-uterine and extra-uterine development of man will be discussed structurally and physiologically. Extended treatment will be given to the study of the adaptation of man and his homeostatic capacity. The physiological discussions will focus on the endocrine organs as a system utilized by man for adaptation to environmental change. Prerequisites: Human Biology 2A and 2B; must be taken concurrently with 3B.

4 units, Win (Kretchmer) MWF 9

3B. Development of Behavior—This course builds on the study of the evolution of behavior in its prerequisite course, 2B. The bases of behavior development are examined in the contemporary human species. Emphasis is on those capacities and behaviors in infancy and childhood which are adaptive for the individual and the social group. The emotional, social, and cognitive development of the human child is considered, beginning with the neonatal period. The utilization of this information is discussed for medical, educational and other social institutions. (This course has overlap with Psychology 111 and credit cannot be obtained for both.)

4 units, Win (Siegel, Staff) MWF 10

4A. Biology of Populations—The course will present a systematic approach to populations as biological units; the dynamics of population growth and the control of population size in the nonhuman and human populations. Demographic principles and community ecology will be emphasized. This course will include treatment of the structure of food webs, the flux of energy through communities, the flux of materials, renewable and nonrenewable resources and how these factors relate to the population dynamics of man. Prerequisites: Human Biology 3A and 3B; must be taken concurrently with 4B.

4 units, Spr (Cavalli-Sforza, Ehrlich) MWF 9

4B. Social Organization of Man—This course will present selected economic, cultural, and sociologic principles relevant to contemporary problems of human biology. Certain data and concepts of the social sciences will be considered, and their significance explored in relation to some aspects of health, disease, and other areas where biology and the social sciences interact.

4 units, Spr (Dornbusch, Gurley, Corning) MWF 10

6. Workshop in Human Biology—This workshop, required of all Program majors, offers the student the opportunity to augment his formal course work with a supervised field, community, or laboratory project of his own choosing. To be arranged in advance. Limited to majors in Human Biology. Course graded pass/no credit exclusively.

4 units (Liebes) by arrangement

10. Human Sexuality—Human sexual function and behavior will be reviewed from biological, psychological, and cultural perspectives. In the first part, the anatomy, physiology, and endocrinology of sexual and reproductive functions are examined. The second part deals with psycho-sexual development and patterns of sexual behavior. In the final portion of the course, erotic themes in literature and art are reviewed, and legal and moral aspects of human sexuality examined. The emphasis in the course is on information, not advice.

4 units, Spr (Katchadourian, Lunde, Staff) MWF 11

Advanced Courses

Note: A major in Human Biology is expected to take 30 units of upper division credits in fields related to the natural or physical and the social or behavioral aspects of Human Biology. The courses may be selected from the upper division offerings of the Program, or any appropriate department on the campus. The student must balance the advanced courses so that two-thirds of the units are in either the natural or the social sciences, while one-third are in the other—i.e., two-thirds social and one-third natural; or two-thirds natural and one-third social. The upper division courses should reflect a unity directed toward the ultimate goal of the student. The student's individual design of advanced courses must have approval from an adviser in the Program. At the student's discretion one-half of the upper division courses (15 units) may be taken for pass/no credit.

Students who plan to pursue graduate work in the sciences or social sciences should be aware of admissions requirements for graduate programs and the necessity for early planning of their programs, in order to satisfy the requirements of both the Program and graduate schools.

Advanced courses presented by the Program in Human Biology are open to nonmajors with the proper prerequisites. Human Biology majors will have preference where the number of students must be restricted.

102. Health as Human Ecology—The interplay of environmental, genetic, and social factors that influence health outcomes. Historical epidemiology, contemporary environmental changes, the evolution of parasites and human hosts, the challenges of health research and of preventive medicine, and the dilemmas of value choices involving life and health will be reviewed. (Students interested in the sociology and economics of medical services should see CPM 200.) Prerequisites: Human Biology core or 20 units of Biological Sciences.

4 units, Win (Lederberg) MWF 11

104. Political Processes and Human Biology—Political practitioners and administrative officials drawn from local, state and national government will discuss issues in effecting changes in public policy. Prerequisite: Human Biology core or consent of instructor.

3 units, Aut (Dornbusch, Staff) given 1974–75

106. Man-Made Environment—A course consisting of lectures, discussions, and readings reviewing man's role in shaping his environment. Emphasis will be placed upon the planning factors and processes which act to determine the nature of our cities and communities. The class is limited to 40 students with preference given to Human Biology majors.

3 units, Spr (Collier) Th 2:05–4:15

110. Introduction to Biological Chemistry—This elective course is designed for students of Human Biology who cannot take courses offered by the Departments of Chemistry and Biochemistry. Major topics include biologically important principles of physical chemistry, characteristics of enzymes and other molecules of biological interest, biochemical pathways and genetic errors of metabolism. This course will be accepted as an advanced course for those Human Biology majors who do not intend to emphasize the natural sciences. Limited to 30 students with preference to students in the Human Biology Program.

3 units, Aut (Hoogenraad) TTh 4:15

120. Human Nutrition—An introduction to human nutrition including the metabolic basis of nutritional requirements, dietary requirements, biogeographic aspects, food production and distribution, specific deficiency diseases, and global aspects of malnutrition. Prerequisite: Human Biology core or consent of instructor.

4 units, Spr (Kretchmer, Robertson) MWF 4:15

128. Seminar in Comparative Survival Strategies—(Same as Political Science 128C.) A systematic comparison of how different human societies go about meeting their basic survival and reproductive needs. The overall configuration of survival strategies and behaviors will be considered, and students will be required to take a holistic approach, drawing upon data and research from a variety of disciplines.

5 units, Win (Corning) T 2:15–4:05

130. Human Genetics—This course will include the following: molecular aspects of human variation, genetics of disease and of continuous traits including behavioral attributes; population aspects; dynamics of change and equilibria under mutation, selection, drift, migration and population structure models; social aspects of human genetics. Prerequisite: Human Biology core or consent of instructor.

4 units, Win (Cavalli-Sforza) MWF 2:15

135. Subsistence Transformations in Prehistory—The course will examine the economic and cultural changes that occurred over most of the Old World during the last 10,000 years. Archaeological and biological lines of evidence will be used to consider the life ways of late hunter-gatherers and the emergence of early forms of agriculture. Discussion will also focus on the major impact that this economic transition has had upon cultural development, human demography, human genetics, and man's relation to the environment. Prerequisite: Human Biology core or consent of instructor.

4 units, Win (Ammerman) MWF 2:15

140. Energy and Society—(Same as Mechanical Engineering 180.) A unified analysis of the effects on man's environment of the production, distribution, and consumption of energy. Treatment will include: the kinds and magnitude of energy resources; the various technologies for conversion to electric energy and other consumer forms; priorities and strategies for future development; the social conflicts between growing demands and environmental degradation; technological assessment; the legal and economic framework of the energy industry. Presentation of technical information will be in terms understandable to the non-engineering student. Prerequisites: high school physics and junior standing or consent of instructor.

3 units, Spr (Connolly, Liebes) TTh 1:15–2:30

150. Biosocial Aspects of Birth Control—(Same as Chemistry 130.) The problems of introducing a new, practical birth control agent or procedure involve legal, political, cultural and economic factors in addition to purely biological ones. The course will deal with a critical evaluation of the logistic aspects of practical human fertility control. Groups of 5 to 8 students of diverse backgrounds will develop a series of position papers dealing with new birth control procedures suitable for populations of different cultural and socioeconomic backgrounds. The selection of students admitted to this class will be based in part on the desire to create a multi-disciplinary student group so that each position paper will be prepared by task forces consisting of participants with different undergraduate backgrounds (e.g., Premedicine, Prelaw, Biological Sciences, Anthropology, Chemistry, etc.) who will focus on specific logistic aspects of a common topic in the birth control field. Limited to 40 students. Preregistration during fall quarter essential. Prerequisite: at least junior standing and completion of pre-registration questionnaire available from Human Biology Office.

5 units, Win (Djerassi) M 2:15–4:05; W 2:15–4:05

160. Primate Behavior—This course will study in detail the research literature on behavior of higher primates in natural habitats. Special attention will be given to chimpanzee behavior, but material on other species of great apes and Old World monkeys will be considered. Some evidence will be included on experimental analysis of questions arising from observation in natural habitats. Prerequisites: Human Biology 2A and 2B.

3 units, Spr (Goodall, Hamburg) by arrangement

161. Primate Behavior Workshop—An African elective; minimum 2 quarters. Prerequisite: Human Biology 160; limited to 8 Human Biology majors per year.

15 units, Aut, Win, Spr, Sum (Goodall) by arrangement

163. Topics in Psychobiology—This course will focus on recent developments in psychopharmacology, as they relate to the study of human mood disorders and schizophrenia. Current theories regarding the etiology of mental illness will be discussed. The relationship between hormones and human behavior will be examined. Emphasis in the course will be on student participation, using a seminar format. Limited to 24 students. Prerequisite: Human Biology core.

3 units, Aut (Brodie) T 3:15–5:05

164. Human Aggressiveness—This course, taught in seminar format, will review data and theory concerning biological, psychological and social aspects of human aggressive behavior. Biological aspects will include instinct theories, genetic variables, hormonal contributions, evidence from the study of nonhuman primates, and brain mechanisms. From a psychological viewpoint, links between frustration and aggression, as well as social learning of aggressive behavior will be reviewed. Social factors will include effects of crowding, stranger contact, status conflicts, and intergroup competition. An effort will be made to integrate information and ideas from biological and psychosocial perspectives. Limited to 20 students. Prerequisite: Completion of Human Biology core program.

3 units, Win (Hamburg, Adams) W 1:15–4:05

199. Directed Reading/Special Projects—Course graded pass/no credit exclusively.

Any quarter (Staff) by arrangement

The Accelerated Medical Program
at Northwestern University:
What Constitutes Adequate Preparation?

ARTHUR VEIS

The ultimate test of the validity of any educational program is the performance of the product, which must be evaluated primarily in terms of specific program goals rather than less well-defined societal goals and values. It is in this sense that I will discuss Northwestern University's accelerated six-year program, and the way we have structured our curriculum to meet some clearly stated goals. I will then try to evaluate our progress in reaching these objectives.

The first paragraph of the formal statement of the goals of the medical school stipulates that: "Students . . . should be well-grounded clinically and sufficiently informed scientifically to grasp fundamental data related to diseases in man. They should know how these data are generated and be able to convert them to useable facts. . . ." In practical terms these objectives can be translated into a strong statement about the importance of the basic sciences related to medicine. Reflecting this, the medical school curriculum retains the first two years of basic science courses, in addition to some clinical work in courses such as physical diagnosis, clinical correlations, and introduction to clinical medicine; the latter two run through the entire first two years.

The Medical Curriculum

I would like to elaborate for a moment on the details of the medical curriculum, since it sets the requirement for our beginning students. In the first year, when we concentrate on the most fundamental types of basic science information, heaviest emphasis is given to the anatomy and biochemistry sequences. Anatomy presents three courses: a fairly classical gross anatomy; microscopic anatomy—in which an attempt is made to bring modern biological and biochemical perspectives to histology; and neuroscience—an interdisciplinary presentation of neuroanatomy in

113

the context of neurophysiological and neurochemical principles. The latter courses require a solid basis in biochemistry and a good background in biology; the neuroscience course has a strong clinical orientation, and clinical correlations are made at frequent intervals.

The biochemistry offering is in two parts: a basic course of thirty lecture hours, and a physiological biochemistry course of thirty-two lecture hours and forty-four laboratory hours which also has strong clinical correlation overtones. Because of the varying backgrounds of the freshmen students, a screening or placement examination is given, and those doing very well, from 10 to 15 per cent, are excused from the basic course but are required to undertake independent study or research. All students are required to take physiological biochemistry.

The balance of the first year is devoted to physiology, genetics, microbiology and behavioral science. The subject matter treatment in all these courses, except behavioral science, leans heavily on the initial biochemistry and anatomy offerings.

I have described this curriculum, which I presume is fairly standard, to lend substance to the statement that we require our students to have strong backgrounds, particularly in chemistry and biology. Moreover, it is our feeling that fundamental courses for medical students, such as physiological biochemistry and neurosciences, are best taught in the medical school. Basic biochemistry and even gross anatomy could conceivably be part of an undergraduate program, although one might seriously argue this point.

The Six-Year Program

Now we turn to the six-year program, which operates as follows: We select a group of sixty students from the top ranks of current high school graduates, and they spend two years in the College of Arts and Sciences on the Evanston undergraduate campus. Half of their program is in science course preparation: one year of physics, two of chemistry, and one of biology. The physics course is a special sequence moving at a fast pace at a relatively high level; chemistry begins with one quarter of the more usual one-year general chemistry material for six-year-program students and others with some advanced placement credit in chemistry. Two quarters of organic chemistry complete the first-year chemistry sequence and bring the students to the level in chemistry achieved by the regular premedical student in two years.

In the second year we require two quarters of physical chemistry, followed by a physical and analytical chemistry course emphasizing laboratory techniques. The latter course was designed specifically for the six-year program but is now open to other students with an interest in biochemistry and biology. Concurrent with the physical chemistry sequence our students must take one year, that is, three courses, in biology. Although a basic sequence is available, with the permission of the instructor we permit those with special interests or particularly strong backgrounds to take more advanced courses. The biology courses are the equivalent of, or somewhat better than, the biology preparation of the regular four-year, nonbiology major, premedical student.

Although these science courses are taken under heavier load conditions than the usual premedical student assumes, the preparation is obviously sufficient. The six-year students have done well in their basic science courses in the medical school. In the medical school biochemistry course in the fall of 1973, for example, twelve of the twenty-three students to receive honors grades were in the six-year program, although students in the program represented only one-third of the class; no six-year-program students failed.

Evaluation of the Program

Looking at the question of successful completion of medical school on the larger scale, and including the clinical areas, the following statistics may be of some interest. The program has been in existence since 1961, and in the intervening twelve years it has enrolled 535 students. One hundred and sixty have graduated with the M.D. degree; 309 remain in residence; and sixty-six have dropped out of the program. Thirty-two students have taken an extra year of liberal arts; six have taken two extra years. Twenty-six participated in the combined M.D.-Ph.D. program; seven have both the M.D. and Ph.D.; eight completed only the M.D.; two stopped at the M.D.-M.Sc. level; and twelve are still enrolled in medical and/or graduate programs.

Of the sixty-six who dropped out, forty-six left during the first half of the program. Thirty of the sixty-six were let go because of poor scholarship; fourteen left because of a change of interest; nineteen for personal, financial, or health reasons; and three died. Seventy to 80 per cent of all attrition takes place in the first two years.

Upon completion of the two years in Evanston, our policy is to enroll the six-year students directly in the medical school without making any kind of identification or distinction. They are thus in direct competition with regular four-year B.Sc. degree students. As I have indicated, few academic problems arise in the basic science years; in our experience there are equally few problems in the clinical years. Aside from an occasional comment by a clinical preceptor on the "immaturity" of some junior clerk, we have not observed any difficulties, qualitatively or quantitatively, experienced by the remaining student peer group. We take this to mean that our six-year students are sufficiently mature and emotionally stable to cope with the rigors of a medical education, and to suggest that they have a broad enough base in general, nontechnical areas to prevent any particularly narrow approach to medicine and life in general. We attempt to insure this breadth during the liberal arts phase of the program.

In our selection process for the six-year program we choose students of exceptional academic ability and demonstrated, wide-ranging interests. Many have advanced placement credit in several subjects. Because of this unusual breadth our only fixed liberal arts requirement is that each quarter the students must take two liberal arts courses in addition to their two required science courses. Students may choose advanced level courses—for juniors and seniors—in accordance with their own interests.

Our policy is to guarantee admission to medical school pending satisfactory performance, which we define in terms of a C or better grade in all courses. By removing the threat of a high grade average and the competitive nature of most premedical programs, we encourage the students to select the most challenging courses and to study in the true interests of scholarship. Our students take advantage of this, and in most cases rise to the challenge this freedom provides.

During the undergraduate phase, we reinforce their commitment to medicine by arranging clinical seminars and regular social gatherings with medical school faculty and students.

Returning to my original theme, I believe that in the Northwestern context the six-year program does prepare students for their medical studies and subsequent clinical careers. They are sufficiently well-grounded in the sciences, for example, to successfully combine M.D.-Ph.D. programs if they wish to do so. The majority of course go directly on to practice clinical medicine and seem to be able to pursue any specialty without difficulty.

Can we recommend such programs as a general approach for other medical schools? Personally, I feel that our success is due to the intrinsic high quality of the students we attract, and to the large available pool from which we can select. Every time we have compromised on high quality and a strong science background, the student has had difficulty in the undergraduate course or in completing the medical curriculum. If all schools adopted a six-year program, the programs could not demand the high level of achievement we now require. If training in medicine requires a strong science base, as we think it does, then the six-year approach as we practice it cannot be generally applied. For a certain group of students, perhaps ten times as many as we can take, such programs do work. The strongest proof of this is that the graduates of our program are the strongest proselytizers for new candidates—including their younger brothers and sisters.

General Discussion

The number of students in the six-year program at Northwestern was recently increased from forty to sixty. Three to 4 per cent of the medical students have come out of early admissions programs or six-year continuum programs. Early admission programs may, however, militate against colleges and universities that do not have associated medical schools.

The program is now fully financed by the university and receives no outside support. Student tuition fees at the Evanston campus and at the medical school are the same for all six years of the program.

Students are actively discouraged from taking science courses during their elective periods in the first two years of the Northwestern program.

Minority group students are enrolled in the program, but some from inner city schools require longer schedules because of inferior science backgrounds.

Education in Human Genetics:
The Program at Yale University

LEON E. ROSENBERG

When I entered medical school in 1953 the teaching of human genetics was no problem—with very few exceptions it simply was not taught. This omission did not of course reflect a total lack of information about man's heredity. Fifty years earlier, Garrod, in his classic study, "Inborn Errors of Metabolism," had been instrumental in rediscovering Mendel's laws and had proposed that genes act through the synthesis of enzymes. Between 1910 and 1930 the mathematical principles governing the appearance, maintenance, and equilibrium of certain alleles in human populations were enunciated by Hardy, Weinberg, Haldane, Fisher, and Wright. In the 1940s the genetic basis for the difference between sickle cell anemia and the sickle cell trait was proposed by Neel and Beet, and soon thereafter this information led Pauling, Itano, and their colleagues to their brilliant conceptualization of "molecular disease."

Why then had these seminal contributions and the large body of information about the genetics of bacteria, neurospora, plants, and drosophila not yet led to the development of a scientific discipline ready to take its place alongside the other basic medical sciences? I believe there were several reasons. First, the chemical nature of genes and their cellular organization were unknown; second, the universality of the principles of genetics in the biological kingdom was unrecognized; third, the impact of hereditary factors on disease manifestations was unappreciated; and, fourth, the Flexnerian orthodoxy that had proclaimed anatomy, biochemistry, pharmacology, and pathology as the basic medical sciences had gone unchallenged.

In the past twenty years all this has changed. The chemistry and molecular biology of the gene and the dogma of protein synthesis are taught in elementary school. The notion that what is true for *E. coli* is also true for the elephant and for man is no longer a subject of debate. The existence of catalogs containing more than 1,600 human Mendelian phenotypes, and the evidence that hereditary factors are a major cause of mortality and morbidity in fetuses, children, and adults, attest to the need for physicians to understand genetics. These developments have in fact come about so rapidly that they have taxed university officials, medical school administrators, and faculty members, as well as curricula.

Varied Patterns of Instruction

I believe it is fair to say that there are as many patterns for teaching human genetics in medical schools in the United States as there are medical institutions. Some schools still have no discrete course in genetics and depend on departments such as biochemistry, microbiology, and epidemiology to present genetic principles. A small but growing number of schools have created separate departments of genetics, human genetics, or medical genetics, and have delegated to them responsibility for a formal course to be taught with the other basic medical sciences in the first two years of the curriculum. Still other schools depend on clinical departments such as medicine and pediatrics to organize sections devoted to teaching the theory and practice of medical genetics in the second, third, or fourth years of the undergraduate medical curriculum. This interinstitutional variability is accentuated by large differences in the number of hours of instruction, and in the "required" versus "elective" notation of the genetics offering. Since there is no current data or planned study that can tell us which of these approaches is most successful, I will illustrate these general remarks by an anecdotal description of the changing pattern in the teaching of human genetics at one institution, the Yale University School of Medicine.

The Yale Program

Prior to 1965 no discrete faculty grouping or course existed. In that year a Section of Medical Genetics staffed by two junior faculty members was established in the Department of Medicine and was asked to present an eight-hour block of lectures on human development in the first-year undergraduate medical course. By 1968 the faculty of this section had grown to four, the organization had become an interdepartmental division sponsored jointly by pediatrics and medicine, and the lecture hours in human development had increased to twelve.

In 1969–70 President Kingman Brewster appointed an advisory committee composed of senior faculty members from the medical and graduate schools to make recommendations about the future organization of research and teaching in the field of genetics. Based on a perceived need to increase faculty representation in certain subdisciplines of genetics, and to promote interaction between geneticists in preclinical and clinical departments, the committee recommended that a Department of Human Genetics be established in the medical school. This department, unlike any of its predecessors at Yale, was not to be organized along "preclinical" or "clinical"—or "basic" and "applied"—lines. Rather it was to be charged with responsibility for teaching genetics in all four years of the undergraduate medical curriculum; fostering collaborative research aimed at solving inherited human disease problems; and implementing a Ph.D. program with emphasis on the genetics of higher eukaryotic systems. Its faculty, reflecting this purposeful blurring of traditional responsibilities, was to be composed of Ph.D.'s and M.D.'s whose combined research interests spanned the field from molecular to clinical genetics.

The Department of Human Genetics was launched in July 1972, and in the past eighteen months we have begun to implement its bold and exciting charges. Considerable progress has been made: ten primary faculty positions have been created by transferring individuals from pre-existing departments: medicine, pediatrics, and microbiology; five joint appointments have been made to scholars from other university departments, and two new faculty members have been recruited; space needed to bring the departmental faculty into contiguity has been identified; and a Ph.D. program has been launched and has enrolled its first three students. The department has added to its other medical school offerings an eighteen-hour course in human genetics in the first year of the medical curriculum; a twenty-two-hour course on biochemical mechanisms of inherited disease, taught jointly with the Department of Molecular Biophysics and Biochemistry; a six-week, full-time clinical genetics elective in the third or fourth years; blocks of four to six lectures on growth and development and biology of disease in the optional third-year combined tracks; and selected lectures in such courses as pharmacology, introduction to medicine, and epidemiology.

Some Obstacles to Progress

Many problems, expected and unexpected, have been encountered. The federal government's decision to cut back research funds and to curtail training grants has affected us, directly, by reducing financial support for faculty, postdoctoral fellows, and graduate students, and, indirectly, by forcing other departments to guard more jealously their budgetary and spatial prerogatives. The absence of contiguous departmental laboratory, teaching, and administrative space has hampered collaborative research efforts. The philosophical gulfs within our faculty engendered by prior departmental memberships, prejudices concerning priorities for faculty recruitment, and mutual ignorance on the part of one group of faculty about the talents and aspirations of the other—that is, Ph.D.'s versus M.D.'s—have closed slowly. Finally, the department has been led by an individual whose skills, such as they may be, did not include traversing the maze of tough university politics.

I cannot assure you that we have adopted the best model for an educational experiment in human genetics, or that we will have the wisdom and courage to test it convincingly. I do believe, however, that the discipline of human genetics contains a sufficient body of common information to span the distance from the undergraduate classroom to the practicing physician's office. Since that span includes people and precepts vital to an educated society's view of itself and its medical needs, it must indeed be bridged.

General Discussion

THEODORE T. PUCK

It is impressive that similar developments in human genetics are in progress in two such widely separated schools as Yale and Colorado; this would seem to indicate the existence of broad principles that apply to programs in any institutions.

Human genetics is too all-encompassing a field to be lodged in a single basic science or clinical department.

In addition to a basic required course in the core curriculum in the sophomore year, the Colorado program includes electives in cytogenetics, laboratory exercises in somatic cell biology, and macromolecular techniques. The graduate program leads to the Ph.D. degree in biophysics or genetics and the combined M.D.-Ph.D. degrees; and special research opportunities are offered to medical students, interns, residents, and postdoctoral fellows.

There are clinical programs in genetic counseling and amniocentesis conducted jointly with the departments of pediatrics and obstetrics-gynecology. A new community service in genetic medicine serves outlying areas.

The Three-Year Medical Curriculum
at McMaster University

VICTOR R. NEUFELD

In this presentation I propose, first, to review the history, philosophy, and curriculum outline of the three-year undergraduate medical program at McMaster; second, to discuss the role of the basic biological sciences, particularly as it is reflected in the organizational structure; and, finally, to describe a recent event that illustrates something of the evolving nature of the program and of the people involved.

History, Philosophy, and Curriculum Outline

Let me begin by reminding you of the relatively short history of the medical program (Figure 1). John R. Evans was appointed as dean in 1965, and by 1966 the first few "founding fathers," as we now call them, had been recruited. The first class of twenty students did not arrive until 1969. As a more recent participant in the McMaster adventure, I have become increasingly aware of the importance of that three-year planning period, from 1966 to 1969, when the philosophical approach was conceived and nurtured.

The sources of the concepts included perspectives from current problems in the health care system and from general education, and researched information about the factors that enhance learning. The overall principle was the emphasis on *learning, not* on teaching. During the planning period, also, many of the original faculty members worked together in interdisciplinary groups, hammering out objectives and preparing programs and learning resources. All this happened before the first students were enrolled.

The class size has grown steadily to the current total of eighty. Two classes have now graduated from the program, the first of these concurrent with the opening of the McMaster Health Sciences Centre.

The philosophical approach consists of at least six basic elements, three of which I have called primary approaches to learning (Table 1).

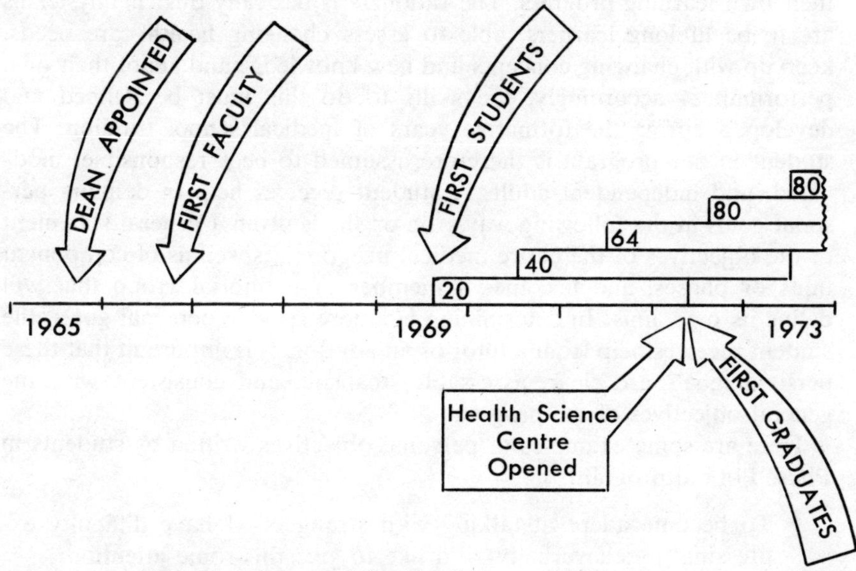

Figure 1. History of the McMaster medical program

My descriptions of each component will be brief, and I will use examples where appropriate.

Self-Directed Learning

The first idea, *self-directed learning,* involves helping students define their learning needs, select appropriate methods of learning, and evaluate

TABLE 1. BASIC ELEMENTS OF THE MCMASTER MEDICAL PROGRAM

Primary:

Self-Directed Learning
Problem-Based Learning
Small-Group Learning

Secondary:

Selection of Learning Resources
Integration
"Formative" Evaluation

their own learning progress. The rationale is basically this: if physicians are to be lifelong learners, able to assess changing health care needs, keep up with changing concepts and new knowledge and adapt their own performances accordingly, the skills to do this must be defined and developed during the formative years of medical school training. The student in our program is therefore assumed to be a responsible, motivated, and independent adult. A student receives help in defining personal goals in the following ways: he or she is given a general statement of the objectives of the entire medical program, as well as of component units or phases, and becomes a member of a tutorial group that will define its own aims. In determining his more specific personal goals, the student receives help from a tutor or an advisor. It is important that these personal goals are clear, assessable, realistic, and consistent with the general objectives of the program.

Here are some examples of personal objectives written by students in Phase I of our program:

- To become adept at talking with strangers—I have difficulty expressing myself verbally—I'd like to give this some attention;
- To get some idea of the workings and effectiveness of the Canadian health care system;
- To try to integrate the study of regional anatomy with the various Phase I units;
- To develop greater proficiency in the use of medical library facilities.

The schedule then allows the student to "do his own thing." Very few lectures are offered to the class: a maximum of one a day, and many days none at all; even so, such large group events are optional. A wide range of learning resources are available to the student to help him achieve his objectives.

Just as the determination of goals and the selection of learning experiences are the student's responsibility, so is self-evaluation part of being a self-directed learner. The prerequisites for the latter are: the student's willingness; an awareness of his goals; and some idea of the expected level of performance or learning. This is a constant and informal process. The student gets feedback about his performance in discussions with his peers; in tutorial sessions; and, rather more formally, in reviewing a written statement of a problem with his tutor, or engaging in self-assessment exercises of various kinds. At the end of each unit, he reviews his progress comprehensively with his tutor.

Problem-Based Learning

The second element, *problem-based learning,* represents an alternative to studying blocks of classified materials in a strictly organized sequence. Since the problems encountered in medicine are primarily those of individual patients, most problem situations presented to our students relate to individual clinical cases. There are many advantages to this kind of learning: it encourages motivation; it stimulates active intellectual processes at the higher cognitive levels; it probably enhances retention and transfer of information; it can be modified to meet individual student needs; and it promotes curiosity and systematic thinking. To the extent that the problems presented are representative of a spectrum of clinical cases, problem-based learning is relevant to clinical practice. But it consists of more than simply learning around clinical problems: it represents a fundamental intellectual process that can be applied to physiological problems in research laboratories, to problems of family dysfunction, and to issues relating to health care in the community.

Perhaps an example may help. Here is a brief case description that was part of a recently completed Phase I program:

Sheila Farkis is six. She is the third child of healthy parents. However, labor was prolonged at the time of her birth. Her two older siblings are in good health. She was noted to be slow in her early motor milestones but in social and language development was considered normal. At 1½ years of age she was assessed and found to have an ataxic form of cerebral palsy (inability to smoothly coordinate her movement and balance). She attended the Cerebral Palsy Centre from then until 4 years of age at which time she went into a cooperative nursery. The following year she entered kindergarten. In kindergarten Sheila was progressing fairly well but the teacher was worried because she felt that Sheila would be at risk because of her physical awkwardness, and saw her as a very vulnerable child. Because of her motor incoordination, Sheila also had trouble writing and carrying on other eye-hand activities such as cutting with scissors. As a result, her work tended to be untidy and sloppy. Because of her poor hand skills and her physical vulnerability, the teacher suggested that she go into some protected class situation specifically designed for handicapped children.

The parents are unhappy with this decision, preferring to see her in a regular Grade 1 class. In stressing their desire that Sheila be treated as normally as possible, they indicate that she is expected to share household chores with her sibs, who become angry if Sheila doesn't "do her fair share."

Each tutorial group, consisting of five students and a tutor, would address themselves to this problem and begin asking *questions*. Arising out of the question would be underlying *issues* such as:

- The relation between prolonged labor and subsequently recognized cerebral palsy;
- The anatomic localization of the lesion of cerebral palsy;
- The normally anticipated "motor milestones" in a child's development;
- The mechanisms of coordinated movement and balance, with the example of ataxia as a form of altered coordination;
- The emotional development of a child with a normal intelligence but abnormal motor function;
- The school system and its response to handicapped children;
- The definition of "handicapped": Who thinks this child is handicapped? Apparently her siblings did not, but her teachers did. Who is right?

The list of issues may be quite extensive, so the tutorial group would decide on priorities. The selected items would be explored within the group as far as possible, without advance preparation; questions requiring further research would then be identified, with designated responsibility for looking for the information. At a subsequent tutorial these questions would be resolved and certain principles would be highlighted. The group may then if it wished look at a list of issues the problem planner thought were relevant to this problem, and compare their list with his. On the average, each group would spend three or four days on such an exercize. The learning of the individual student would be directed partly to the objectives determined by the group, and partly to his own needs.

The problems in Phase I were designed to lead to a broad-ranging exploration of universal issues and concepts. In later phases the exploration is more focussed and in greater depth.

Small-Group Learning

With respect to the third element, one of the goals of the program is that a student will become a productive contributor to a small group, be it a learning group, research group, or health care team. The rationale for the *small-group learning* approach is multifaceted.

The small-group tutorial represents a laboratory for learning about human interaction, in which a student can develop interpersonal skills and become aware of his own emotional reactions. It is an opportunity to learn to listen, to receive criticism, and, in turn, to offer constructive criticism. It is a forum for group problem solving, where the pooled resources of the members, in terms of academic training, experience, personality, and perspective, are more effective than the sum of individual abilities. It also offers an opportunity for self-evaluation, in that a student can informally compare his own learning progress with that of his peers.

The faculty tutor has a key role in this learning group. Although he is an "expert," or content-area specialist, in some branch of medical science, his role as a tutor is primarily that of a generalist and facilitator. He must be skilled in managing small-group interaction; he must coordinate effective and meaningful evaluations; and he should himself be an example of self-directed learning and problem solving.

Backing up these three primary approaches are three secondary concepts. Given the variation in student needs and learning styles, a wide variety of learning resources are available, some of which are most useful for stimulating problem solving. These include patients, actual or simulated, and clinical situations portrayed in various audiovisual formats. Other resources are mainly intended to provide information—selected readings, slide-tape presentations, and resource persons who are available to students or tutorial groups, much as a consultant is available to another physician.

I will expand on the idea of integration later, particularly as it pertains to the basic and clinical sciences.

Finally, we tend to put our energies into evaluation that occurs *during* a learning period rather than at the end: thus evaluation is used for educational diagnosis and prescription, and can shape or form the subsequent learning experience. This kind of assessment occurs in many ways —by self-evaluation; by working in a tutorial group with other colleagues; and by individual discussions with various faculty members.

Before I outline the actual program, let me state the two important principles that guide us in the selection of students. The first is a serious attempt to choose them not only for their academic credentials but for their personal characteristics. We look for such evidence as their demonstrated ability to be independent learners and imaginative problem solvers, their emotional stability, and their potential to contribute productively to a group. The second principle is the selection of students from a wide range of academic backgrounds—close to a half of our students come from premedical experiences other than in the biological sciences.

Because of this second selection policy we arrange a special four-week course for students who have had little or no cell biology or biochemistry. The purpose is to introduce them to the basic vocabulary, and to allay their anxiety about not having a "biological" academic background; the planners of this course firmly believe that the second objective is the more important. There is a probability that this course will be integrated with our first phase in 1974–75.

Overview of the Program

Now I will give a brief overview of the program (Figure 2). It can be seen that the actual duration is somewhat less than three years; that there are four phases; that there is a substantial amount of elective time; and that there is a break in the action in August.

Briefly, Phase I is a multifaceted introduction to the community; to the faculty and facilities of the institution; to the various learning approaches mentioned earlier; to universal concepts of human behavior, structure, and function; and to clinical skills.

Phase II is entitled "Response to Stimuli." Here, five pathophysiological models are explored to see how an individual's homeostatic mechanisms attempt to withstand injury. They are: ischemia, inflammation, reactive depression, metabolic homeostasis, and neoplasia. Phases I and II each last about twelve weeks.

Phase III is organized into four ten-week combined organ system blocks bearing such colorful titles as "Blood and Guts" and "Neuro-

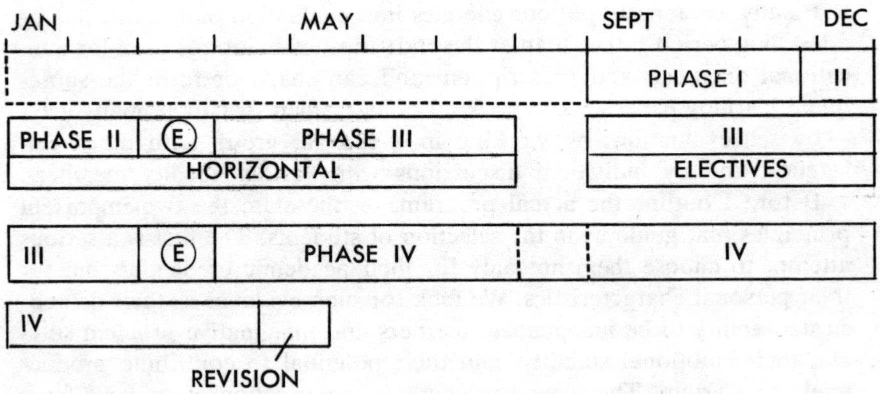

Figure 2. The curriculum outline at McMaster University

Psycho-Loco." In this phase, clinical problems are explored in greater depth, with the emphasis on the physical and behavioral mechanisms that lead to certain clinical manifestations.

Phase IV, the clinical clerkship, currently consists of several combined rotations: medicine-surgery, obstetrics-gynecology-pediatrics, and family medicine-psychiatry. There are several designated elective programs in Phase IV, including the so-called "horizontal electives" that run concurrent with Phases II and III. There are also block electives between these phases, each of which takes about four to six weeks.

Up to this point, in presenting a brief overview of the history, philosophy, and curriculum outline of the McMaster medical program, questions undoubtedly will be raised as to the role of the basic sciences in the program. I would like to anticipate four frequently asked questions.

How Are the Basic Sciences Defined?

The term "basic sciences" is usually considered to include the basic physical sciences such as biochemistry, physiology, and cell biology. One could argue that the behavioral sciences of sociology, psychology, and anthropology are equally basic to the understanding of human health and disease. Kerr White has used the designation, "the new basic sciences" to include epidemiology, biostatistics, and medical economics; and Feinstein has argued for an understanding of the science of clinical judgment.

We have tried at McMaster to incorporate principles from all of these disciplines, whether physical or biological, behavioral, epidemiological, or clinical—all are considered to contribute to the understanding of problems of human health and disease. Throughout the entire three-year program a spectrum of concepts, from the molecular to the social, is kept before the student, with the individual patient as the central focal point.

Is There an Optimum Learning Sequence?

Sequences are arbitrary decisions that are usually unfounded in real life. While it is commonly assumed that the normal should be learned before the abnormal, it is the experience of many that only by comparing the abnormal with the normal can a range of normal become clear to the student. Included in this assumption, also, is the idea that the basic sciences must be learned prior to the clinical sciences. Some programs have reversed this order and expose the students to clinical experiences first, and then bring them back to the basic sciences. Still other programs have clinical and basic sciences running concurrently.

Is There a "Core Content?"

We have rejected the idea that the faculty should devote their energies to defining which facts *must* be learned by the student in order to practice effectively ten years from now. Recognizing that knowledge is tentative, we ask the consultants to specify important current concepts in their disciplines which can be used as guides by unit planners. Unit planners in turn present to the students *general* objectives for a particular unit, representing a synthesis of suggestions from many disciplines. They also present a variety of clinical problem situations to be used as a base for learning.

From this point, each tutorial group and each student works out a study plan incorporating group and personal objectives. This process is crucial because our students are a heterogeneous lot, with half of them, as mentioned earlier, coming from nonbiological academic backgrounds, such as psychology, mathematics, sociology, engineering, and many other fields. For a unit planner to present detailed "core content" objectives for such a mix of students would be inappropriate. The core is determined at the level of the group or individual.

This may sound somewhat "hit-and-miss." Might there not be a risk of major gaps in a student's knowledge at the end of three years? Will he be a "safe" physician? There are undoubtedly gaps in our students' knowledge, just as there are in those from other schools. There are, however, several built-in safeguards that prevent major gaps from developing:

1. The students learn how to identify their weak areas and shore them up.

2. Program planners are advised by "discipline consultants," who in effect are asked to:

    a) Suggest where in the program their concepts could most appropriately be introduced;

    b) Suggest *how* they could be learned; and

    c) Prepare learning resources that can be used when required—key references, slide-tape demonstrations, etc.

3. Dozens of situations are presented to the student throughout the three-year program—variety, relevance, and increasing complexity are the factors considered in selecting the problems.

4. Each unit planner sets general goals as a framework within which each student and group may define more specific objectives.

5. Students are exposed to a variety of disciplines through interaction

with their fellow students and tutors; the tutorial groups are reformed every ten weeks.

6. In addition to problem situations presented by planners, students have contact with actual patients from the beginning of the program, and are encouraged to read about and study their patients' conditions.

7. The clinically relevant gaps are quickly detected and remedied during the clerkship.

What is the evidence that there are no major gaps? First of all there *are* some students who have problems and for whom there is a remedial system that is improving year by year.

Perhaps a more appropriate question is: Are we producing "safe" physicians? While it will be a matter of time before we know, the feedback from internship and residency programs on our first two classes is in general most favorable.

Who Is Responsible for the Educational Program?

The educational program, and subsections of it, are the responsibility of program planning groups composed of faculty members from a broad range of disciplines, along with student participants. An educational program is therefore not a department-based responsibility, and members of the planning groups do not specifically represent their departments or disciplines. They are, however, selected because of their interest and their planning ability.

Any faculty member, including those in the basic biological sciences, may be asked to perform in a variety of educational roles, some of which relate directly to his specialty:

- As a "discipline consultant" or advisor to the planner.
- As a resource person who prepares learning materials or meets on request with students about a question that requires his expertise.
- As elective supervisor.

Other roles unrelated to his specialty are:

- As tutor, perhaps the most important of all the functions, his task is in general not related to his specialty.
- As a unit planner or phase chairman he can be from any discipline.
- As a faculty advisor, again he need not represent his specialty.

The current administrative organization is a result of the desire that the educational program be integrated and interdisciplinary. It was realized early in the program's development that this would require an arrangement where an individual faculty member would be accountable, with respect to his educational contributions, to the director of the educational program rather than to a department chairman. A similar policy emerged in the way research programs were originally organized on a broad interdisciplinary basis. So an administrative model developed in which educational and research programs were seen as goal-oriented activities, and departments were thought of as sources of expertise within a particular discipline.

It was not until this arrangement had been in operation for a year or so that Dean Evans was told that McMaster had reinvented a system that is known in business circles as the "matrix management" model. We have since learned more about this model, and have gradually extended its application to include other educational programs within the Faculty of Health Sciences, as well as research programs, and, to a lesser extent, health service programs. In the research area, for example, several large interdisciplinary research teams are each pursuing a particular theme. All research programs are coordinated by the Committee for Scientific Development, which is responsible for allocating resources, facilitating the development of new programs, and monitoring existing ones. The committee also makes recommendations to the Faculty Council, the body responsible for program priority decisions.

A Recent Adventure

Finally, I would like to describe a recent happening in the program that I believe illustrates the character of the people involved and the development of the ideas that influence our activities.

To set the stage, I should state that at the end of their undergraduate training all Canadian medical graduates take a set of national certifying examinations to qualify for the diploma, Licentiate of the Medical Council of Canada (LMCC). The examinations consist for the most part of Part II of the National Board of Medical Examiners multiple-choice questions, and patient-management problems (PMPs) from Part III of the examinations. Before our first students were enrolled, the faculty debated what the McMaster position would be on these examinations. Most argued that they were not a good measure of what the McMaster program was all about—at best they were an incomplete assessment of factual information in the clinical sciences at a certain point. They certainly did not measure such qualities as self-directed learning, contributions to a learning group, responsibility, self-appraisal, and so on, all of which were the stated goals of the program. It was agreed, however, that passing them was a necessary prerequisite to obtaining a license in Canada, and that our students should not be disadvantaged when it came to taking the examinations.

Consequently, toward the end of the final year of our first and second classes, a revision period of six to eight weeks was set aside so that the students could prepare. In addition, they were given multiple-choice tests that simulated the LMCCs.

The first class passed both sections without difficulty and did quite well. The response of the faculty was mixed; some said the results should not influence our program one way or another; most felt mild relief that we could carry on and "do our thing."

The results of the second class of forty students indicated that one had failed to pass the multiple-choice questions (equivalent to Part II of the national boards), and that five had failed by a narrow margin to pass the section of PMPs. There were a number of reactions to this situation, and perhaps I can best summarize them by stating the implications for: 1) the students; 2) the program; and 3) the LMCCs as a measure of clinical competence.

1. There was a genuine feeling of disappointment among both students and faculty that the program had not fulfilled the objective that our graduates should be able to pass the LMCCs. Five students were let down and we felt at least partly responsible for their failure. It has made us take another look at the learning records of these students—a point I will come back to shortly.

The students currently enrolled reacted in a number of ways: some were predictably anxious; many did not understand how the examinations were designed, scored, and evaluated. They did not realize, for example, that most Canadian schools require their students to take some internal screening examinations prior to taking the LMCCs, and refuse to allow some students to take the LMCCs. All of our students are allowed to take the LMCCs. Virtually all Canadian candidates eventually pass, if not in the first year, then in the second attempt.

Many students realized they had to take their roles as self-evaluators and peer evaluators more seriously. It is commonly accepted that a student's peers are among the best judges of those who need help and in what area. This is certainly true of the present class.

Most students, along with the faculty, cautioned against overreacting to the results. In general the message has been: "Let's learn what we can from this, but let's not change the system on this kind of evidence; let's just make the system work better."

2. With respect to implications for the overall program, we have realized that most of the students who failed had been in academic difficulties of various kinds during their three years in the program. It is also of some interest that we had been more worried about some students who passed the LMCCs than about the five who did not. The difficulties reflected the immaturity of the program in its early stages in terms of the selection process; the functioning of the tutorial groups; and the evaluation system.

I think it is quite clear that many applicants to the program would not thrive in our environment; they require a more structured atmosphere. As we studied the early evaluation system it was apparent that in several cases the tutors did not confront the students with their unsatisfactory progress until the end of a particular unit; they should have done so earlier and provided the help that was needed. Moreover, the faculty advisory and special student assistance systems in their early formative stages were not very effective. There has been steady progress in all of these component parts of the system since that first year.

3. Finally, we are now taking another look at the LMCCs and their significance for the McMaster program, and we have again affirmed the position that our students should be able to pass these examinations. We plan to give the next class more systematic help in the use of such techniques as the PMPs, with which our students have had little experience. We still consider, however, that these measures assess only a small part of what is commonly known as clinical competence. Our efforts to develop new and perhaps more appropriate measures of evaluation will continue unabated.

Summary

In summary, I have tried to present to you the evolving story of a program characterized by the following features:

- A clear statement of general goals and methods;
- An emphasis on the learner, and on the facilitation of learning rather than teaching;
- A balance between the basic biological sciences and the basic behavioral, epidemiological, and clinical sciences;
- An organizational arrangement that promotes interdisciplinary, integrated educational, research, and patient care activities;
- A climate of internal review, candid evaluation, and program modification based on appropriate information.

We have taken some risks and made many mistakes—but enthusiasm and energy levels remain high. And there is a steadily increasing confidence in the central set of ideas, which continue to attract faculty and students to join this adventure.

&

General Discussion

The management approach described by Dr. Neufeld may be more easily applied at a new school such as McMaster than at an established institution. The emphasis on self-learning at McMaster drew special praise: the performance of the school's graduates as interns has been excellent. Five or six students is regarded as the optimum size for a tutorial group. The impression at McMaster is that its tutorial system and the lecture system at the University of Toronto are quite comparable in terms of faculty man hours.

There is a two- to three-month orientation program for tutors, employing workshops, self-study units, and videotapes. Of greater importance, however, are the so-called tutorial training or support groups during that educational period.

The fact that tutors are expected to teach in scientific areas outside of their own specialty is a major benefit; they have to learn to say: "I don't know."

Teaching Biochemistry at the Clinical Level: The Program at Duke University

JAMES B. WYNGAARDEN

I will discuss one experience in teaching biochemistry at the clinical level at the Duke University School of Medicine, which I think has both representative and special features. Before going on, however, I want to be sure my remarks on that more limited topic are understood in the larger context of the total curriculum and environment.

The curricular revision at Duke was conceived in about 1961 and was put into operation five years later. It developed against the background of the Western Reserve experience a decade earlier, and in the context of a limited experiment in medical education—the Cell Biology Research Training Program—that was started at Duke in 1959 as an elective component of the medical curriculum. I was asked to direct the program.

One purpose of the program was to provide a substantial research training experience for medical students and young physicians who were interested in a research-oriented career; another was to recruit additional faculty in areas in which we were not very strong at that time. We added six new basic science faculty members, all of whom came from institutions where they had had no contact with medical students, such as the Rockefeller University and the Biology Department of Columbia University.

In order to make the program available to students without lengthening the total time in medical school, some rearrangements were made in the schedule. Students entered the program after completing about a half-year of clinical work, and, in addition, two summers of clinical electives. This permitted them to spend the third year in research training. The students were self-selected by application, and then by the faculty, so they were not typical. The vigor with which they returned to a second experience in basic science learning after a brief sojourn in the hospital impressed us so much that one of the major objectives of our general curricular revision became the provision of such an opportunity to all students.

By the early 1960s we had become dissatisfied with the rigidity of the curriculum adopted when the school opened in 1930, which provided one-half time free for elective work. Over the subsequent decades it had obeyed the gas laws and had gradually expanded to fill every available space. We felt that was unwise. We also thought that the lockstep nature of the curriculum was inimical to the students' different rates of growth. Our students were no longer interchangeable on entry to the medical school, and they certainly were not interchangeable on exit. We wanted to increase the flexibility, as well as the number of options available to students.

Implementing the New Curriculum

After much planning and a number of faculty retreats our new curriculum was put into effect in 1966. We decided that one-half of the curriculum would be under the general control of the faculty, and that the other half would be elective and under the general control of the students. Having decided that, we rather arbitrarily divided the first two years into two equal components consisting of basic sciences and clinical sciences. Within those two years the apportionment of time was based on the previous allocations in various basic science and clinical areas, with some negotiated exceptions.

The condition that made this revision acceptable to the basic science departments was that students would come back to the basic sciences for one of the last two years. There were to be no specific course requirements; the choices were to be elective, and there would be an advisor system. Thus the basic science departments would have more time than they had had previously, namely, two full years compared with the former one-and-a-half years.

The first year now consists of required courses in basic sciences, and the second year of introductory clinical clerkships, five of eight weeks each, taken in any order. At the end of the first two years the student has finished all formal requirements for graduation, except for two additional years in medical school when he must accumulate about an equal number of basic science and clinical credits.

Basic Science Electives

We knew in advance that basic science electives in the fourth year would not work. Under the original system there had been some elective time in the fourth year, but the students never used it to come back to basic sciences. By that time they were concerned about their internships and wanted to be as competent clinically as they could by the end of that fourth year. The elective basic science year therefore became the third year. As this year has evolved, the basic science electives have taken essentially three forms:

1. Course work offered by virtually all basic science departments.

2. About twelve interdepartmental and interdisciplinary study programs, such as the neuroscience program, developmental biology program, and so on.

3. One-half to a full year of laboratory research. The student may elect such an opportunity simply as a learning experience, or use it to acquire research skills for his future career.

An additional option of the third year is the Cell Biology Research Training Program mentioned earlier. Finally, in the case of our combined degree programs, medical students enter the graduate school during the third year, and eventually return for a fourth year of clinical training consisting of electives offered by all the clinical departments of the medical school. These include clerkships of a general and specialized nature, and courses and seminar work.

Students who enter Duke with advanced placement can do basic science work that gives credit toward the third year rather than the first. Students who enter with substantial science training, with a Ph.D. in a biological science, for example, are exempted from year three.

Role of the Basic Science Faculty

I would like to say a few words about the role of the Duke basic science faculty in this entire curriculum matter. First, basic science faculty members had a major role in planning the curriculum, and continue to serve on many committees, such as the special committees for each of the four years, and the promotions committees. They also participate in student advisory groups.

Basic science members have other teaching roles in the first and third years of medical school; are part of major cooperative research programs carried out jointly with clinical science departments and incorporating trainees at all levels; and have special responsibilities in the M.D.-Ph.D. programs. The basic science chairmen are members of the medical center's executive committee. In short the basic science faculty is an integral part of the life of the institution.

Biochemistry in Clinical Medicine

In the Department of Medicine, while we do not attempt to teach biochemistry in a formal manner, in terms of clinical learning, continual emphasis is placed on the biochemical base of medical science. Biochemistry has become the language of biology, and the identification of human disease is an exercise in applied biology. Every medical student must become familiar with biochemistry as a language, as a body of knowledge and concepts, and as a working scientific method.

The Department of Medicine at Duke is perhaps similar to those at many other large schools. It has 110 full-time faculty members, twenty-five of whom hold joint appointments in the basic sciences. Thus many of the clinical faculty participate in basic science teaching programs.

Although the clinicians have no departmental role in basic science courses, many individual members of the Department of Medicine are visible in the first year. Some give invited lectures in basic science departments, and they are perhaps most active in correlation exercises in the genetics program. From the time the student enters the medical school he sees many living examples of the utility of basic science knowledge in clinical medicine, and is shown models of physicians who use biochemical skills in their professional work.

Many of the clinical faculty take part in rather basic biological research that is disease oriented or motivated; much of it is good science. As one index of this, in 1972–73 twenty-eight papers by members of the Department of Medicine were published in leading biochemical journals —eleven in the *Journal of Biological Chemistry*.

There are 114 medical students per class, but that is just a fraction of the students that the Department of Medicine has contact with each year, for we teach in all four years, especially in the last three. In actuality we have a close relationship with about 300 medical students in any one year.

Eighty interns and residents serve in the department, in addition to about 100 fellows, including twelve or thirteen Ph.D. fellows, and forty physician's assistants in training. Thus over a one-year period our total student body totals over 500.

In his second year in the Department of Medicine the medical student spends most of his time on the wards. Teaching rounds take up two hours a day, and there is a one-hour conference at noon each day under the direction of various subspecialty groups. These exercises place less emphasis on didactic teaching or organized information transfer than in earlier years, and more on incorporating the student into a representative series of learning experiences in which he plays an active role. He is not held responsible for acquiring any defined body of knowledge. He engages in the study and care of patients, and over a period of time his knowledge of medicine expands. In the context of clinical teaching, whenever possible we attempt to interpret disease mechanisms in biochemical terms; this, as we all know, is becoming increasingly possible each year.

Members of the Department of Medicine who hold joint appointments participate actively in third-year basic science electives, particularly in the special studies programs. In the fourth year, clinical rotations cover all the standard general and specialty areas, many of which, in addition to ward experience, cover special seminars in, for example, cardiovascular pharmacology, renal and electrolyte physiology, and biochemical mechanisms of genetic disease. There is continual reenforcement of basic science learning, and a major effort is made to interpret clinical phenomena in terms of basic science mechanisms.

We recruit a variety of faculty members for the department, some of them specifically because of their combined clinical and basic science skills. In those cases the stamp of approval of the basic science departments is essential, both in maintaining the standards of the school and in insuring the career opportunities of the individual.

I think the revised Duke curriculum is at variance with observations that curricular revision is often made selectively at the expense of the basic science departments, and that it may require a much more extensive change in the life style of the basic scientists than of the clinicians. The Duke reforms placed additional requirements on all departments. While basic science departments have had their stresses, the clinical departments have also had to modify their life style. For example, prior to 1966 the Department of Medicine was responsible for about 19 per cent of curriculum time. That figure is now over 25 per cent—we teach over one-fourth of all hours of the medical school curriculum. Previously we were assigned 25 per cent of the hours in the fourth year; under the free elective system about 50 per cent of all fourth-year student hours are now spent in our department.

Thus the new curriculum has significantly expanded our commitments to both basic science and clinical teaching. The department has accepted these responsibilities gladly, and remains enthusiastic about what we have done. I believe our students are equally pleased.

General Discussion

ALFRED P. FISHMAN

An important factor in the success of the Duke curriculum is that the students are *required* to take basic science electives in the third year. If third-year students are *merely strongly urged* to go back to basic science electives, virtually none of them do so, and the basic science faculty becomes disenchanted with the curriculum. During the third year, also, there is a short course in infectious diseases based on clinical medicine. Some of the electives may use patients for demonstration purposes, but otherwise there is no patient contact.

Most students at Duke return primarily to three departments: pathology, physiology, and pharmacology—departments immediately relevant to clinical medicine. Relatively few take anatomy and biochemistry, as they will not study abstract basic science a second time. By and large, the students' objections to the basic science electives at the beginning of the third year have disappeared by the end of the fourth year.

Curricular revisions are usually made at the expense of the basic science departments: the faculty is called on for more major shifts in its life style. At Duke, however, the Department of Medicine has experienced a considerable change in its life style because of a sharp increase in its teaching load.

Participants

GORDON B. AVERY, M.D.
*Professor of Child Health
  and Development*
School of Medicine
George Washington University
Washington, D.C.

GEORGE PACKER BERRY, M.D.
*Special Consultant to
  the President*
Princeton University
Princeton, New Jersey

BARBARA L. BLACKWELL, DR. P.H.
Medical Specialist
Office of Curricular Affairs
School of Medicine
University of California,
  San Francisco
San Francisco, California

JOHN Z. BOWERS, M.D.
President
Josiah Macy, Jr. Foundation

L. THOMPSON BOWLES,
  M.D., PH.D.
*Associate Dean for Curricular
  and Student Affairs*
The George Washington University
  Medical Center
Washington, D.C.

MARY I. BUNTING, PH.D.
*Assistant to the President
  for Special Projects*
Princeton University
Princeton, New Jersey

SAM L. CLARK, JR., M.D.
Professor and Chairman
Department of Anatomy
The Medical School
University of Massachusetts
Worcester, Massachusetts

VINCENT G. DETHIER, PH.D.
Professor of Biology
Princeton University
Princeton, New Jersey

THOMAS M. DEVLIN, PH.D.
Professor and Chairman
Department of Biological
  Chemistry
The Hahnemann Medical College
  and Hospital of Philadelphia
Philadelphia, Pennsylvania

ALFRED P. FISHMAN, M.D.
Director
Cardiovascular-Pulmonary
  Division
Department of Medicine
School of Medicine
University of Pennsylvania
Philadelphia, Pennsylvania

JOSEPH S. GONNELLA, M.D.
Associate Dean
Director of Academic Programs
Jefferson Medical College of
    Thomas Jefferson University
Philadelphia, Pennsylvania

THOMAS HALE HAM, M.D.
Director
Division of Research in
    Medical Education
School of Medicine
Case Western Reserve University
Cleveland, Ohio

NORMAN KRETCHMER,
    M.D., PH.D.
Faber Professor of Pediatrics
Department of Pediatrics
School of Medicine
Stanford University
Stanford, California

CYRUS LEVINTHAL, PH.D.
Professor of Biology
Department of Biological Sciences
Columbia University
New York, New York

MA LIN, PH.D.
Professor and Chairman
Board of Studies in Biochemistry
University Science Centre
The Chinese University
    of Hong Kong
Shatin, N.T., Hong Kong

C. PHILIP MANIRE, PH.D.
Kenan Professor and Chairman
Department of Bacteriology
    and Immunology
School of Medicine
The University of North Carolina
Chapel Hill, North Carolina

CLEMENT L. MARKERT, PH.D.
Professor of Biology
Department of Biology
Yale University
New Haven, Connecticut

PAUL A. MARKS, M.D.
*Professor of Human Genetics and
    Development*
Vice President for Health Sciences
College of Physicians and Surgeons
Columbia University
New York, New York

ROBERT S. MARSHALL, PH.D.
*Assistant Director for Academic
    Affairs*
Division of Biological Sciences
Cornell University
Ithaca, New York

J. D. MYERS, M.D.
University Professor of Medicine
Department of Medicine
School of Medicine
University of Pittsburgh
Pittsburgh, Pennsylvania

DAVID L. NANNEY, PH.D.
Professor of Zoology
Biology Programs Officer
School of Life Sciences
University of Illinois at
 Urbana—Champaign
Urbana, Illinois

VICTOR R. NEUFELD, M.D.
Coordinator
Program for Educational
 Development
Faculty of Medicine
McMaster University
Hamilton, Ontario, Canada

CHARLES E. OXNARD,
 M.B., CH.B., PH.D.
Professor of Anatomy,
 Anthropology, and Evolutionary
 Biology
Master
Biological Sciences
 Collegiate Division
The University of Chicago
Chicago, Illinois

THEODORE T. PUCK, PH.D.
Director
Eleanor Roosevelt Institute for
 Cancer Research
Professor of Biophysics
 and Genetics
The Medical Center
University of Colorado
Denver, Colorado

THOMAS B. ROOS, PH.D.
Professor of Biology
Department of Biological Sciences
Dartmouth College
Hanover, New Hampshire

LEON E. ROSENBERG, M.D.
Professor and Chairman
Department of Human Genetics
School of Medicine
Yale University
New Haven, Connecticut

GEORGE STREISINGER, PH.D.
Professor of Biology
Institute of Molecular Biology
University of Oregon
Eugene, Oregon

ARTHUR VEIS, PH.D.
Professor of Biochemistry
Associate Dean of the Medical
 and Graduate Schools
The Medical and Dental Schools
Northwestern University
Chicago, Illinois

THOMAS G. WEGMANN, PH.D.
Associate Professor of Biology
The Biological Laboratories
Harvard University
Cambridge, Massachusetts

JAMES B. WYNGAARDEN, M.D.
Professor and Chairman
Department of Medicine
School of Medicine
Duke University
Durham, North Carolina

INDEX